Theory and Applications of Sequential Nonparametrics

PRANAB KUMAR SEN

University of North Carolina

SOCIETY for INDUSTRIAL and APPLIED MATHEMATICS · 1985

PHILADELPHIA PENNSYLVANIA 19103

Library of Congress Card Number 84-52332
ISBN 0-89871-051-0

Filmset in Northern Ireland by The Universities Press (Belfast) Ltd.

Contents

Chapter 6
NONPARAMETRICS OF SEQUENTIAL TESTS

Preface

These lecture notes follow closely the actual presentation at the CBMS-NSF Regional Conference in the Mathematical Sciences at the University of Iowa, Iowa City, July 18–22, 1983. Professor Robert V. Hogg indeed did a splendid job in organization and in providing a stimulating and congenial atmosphere for the conference (despite the "hot diggity days" in July with Iowa temperature consistently above the 100°F mark). All the participants of this conference will remember long the cordial hospitality of Bob (and Carol) Hogg and their colleagues. The excellent set of notes taken by Peter Wollen during the conference has made my task easier in preparing these notes. More than expected typing assistance from Ms. Page S. Boyette is also thankfully acknowledged.

This set of ten lecture notes covers a prime portion of the material (mostly, in Part 2) of my book, *Sequential Nonparametrics: Invariance Principles and Statistical Inference* (John Wiley, New York: 1981), along with some more recent developments and some additional topics not reported in this book. Occasionally, the treatments are somewhat different too, and more emphasis is laid here on the statistical interpretations and applications. These lecture notes have benefited greatly from the sharp cross-examination and critical comments from the participants of this conference. In the following, I have tried to summarize the main theme of my presentations.

> When you can't achieve a true sense of perfection,
>   'to err is human' may bestow you some satisfaction.
> When a statistician can't get rid of a limitation,
>   the only avenue unblocked is the Sequentialization.
> Only when you know the underlying model well, try parametrics,
>   otherwise, surrender unconditionally to the Nonparametrics.
> While 'bounded influence curves' appear only for a local departure,
>   remember, in nonparametrics, 'global robustness' dominates the picture.
> L-, M- and R-procedures are almost the same if you know the distribution,
>   but, in practice, can you ever confide on such a stringent assumption!
> Are the M-statistics scale-equivariant or distribution-free?
>   what about the scale-equivariant L-, when you aspire to be F-free!
> Though the 'maximal invariants' lead to the R-, generally distribution-free?
>   but, can you choose an appropriate one to set yourself limitation-free?
> ASN, OC, risk, deficiency and asymptotic normality of stopping times,
>   are undoubtedly the sequential superstars in the mathematical skies.
> But, what about a repeated significance test when you have a jackknife,
>   a change-point problem or a time-sequential procedure on human life!
> In this conference, these provide us with the contextual bread and butter,
>   and I plan to proceed sequentially to utter something without a computer.

Pranab Kumar Sen
Chapel Hill, North Carolina

CHAPTER 1

Introduction and General Objectives

Our principal objective is to present a systematic account of some recent developments in the methodology of sequential nonparametrics and to focus on potential applications in some live problems arising in clinical trials, life testing experimentations, survival analysis, classical sequential analysis and other areas of biostatistics and applied statistics. Most of these developments have taken place only in the recent past, and the theory is largely of asymptotic character. The asymptotics in this context not only induce a lot of simplifications and unification in the general statistical formulations, but also provide solutions which are adaptable in practice under quite general regularity conditions. Unlike their parametric counterparts, these nonparametric procedures do not require the (usual) assumption that the underlying probability distribution is of specified (functional) form (involving some (usually unknown) parameters), so that the theory remains valid and applicable for a general class of underlying distributions. Or, in other words, these nonparametric procedures have greater *validity-robustness* (in the global sense), and, at the same time, they possess good *efficiency* properties for some important (and typical) subclasses of distributions.

Various aspects of nonparametric sequential procedures (for a variety of statistical inference problems) have been considered in my recent book, *Sequential Nonparametrics: Invariance Principles and Statistical Inference* (John Wiley, New York, 1981). In these developments, a unified (*martingale-reverse martingale*) approach has been incorporated in the crystallization of the *asymptotic theory* which basically rests on some (weak as well as strong) *invariance principles* for various nonparametric statistics and (strong) *consistency* properties of some estimators of their dispersions. Some *uniform integrability* results for these statistics are also needed in some specific problems, and these were studied too. Here, we intend to present a general review of these basic mathematical results (mostly, omitting the details of the proofs) and to incorporate these results in the motivation and formulation of some nonparametric sequential procedures. We shall also put stress on their live applications. In this way, a substantial portion of the material of that book (mostly, Part 2) will be recapitulated here along with some other topics, which were left out there, and some further developments of more recent origins. In some cases, the mathematical treatments are somewhat different too.

In clinical trials and follow-up studies, nonparametric procedures are becoming increasingly popular. In Chapter 2, nonparametrics of these *time-sequential procedures* are discussed in detail (for the simple as well as the multiple regression models and the analysis of covariance problem). In this context, special attention has been paid to the so-called *progressive censoring schemes*.

1

In this setup, one does not need the so–called *proportional hazard models* (which are only quasi-nonparametric) or models relating to specific parametric families of distributions. The nonparametric procedures considered are quite generally adaptable in comparative clinical trials, and in some of the actual adaptations they have been observed to be quite robust and efficient too. Martingale central limit theorems and invariance principles play the key role in this context. The end products are quite comparable to the ones in the classical sequential analysis. Areas of potential applications of this theory of time-sequential nonparametrics include comparative (possibly, multi-center) clinical trials, survival analysis, life testing experiments and follow-up studies. In such studies, one may also encounter *staggered entry* and *withdrawal* of the units from the scheme, and, the necessary modifications to adjust for such complications are also discussed. In terms of the scope of applications in live problems, time-sequential nonparametrics constitute one of the major areas of applicable sequential nonparametrics.

An important problem in industrial quality control is to detect changes in the distribution of sequentially observed independent random variables. A *change-point model* arises typically in this context, though it may also arise in other follow-up studies. This model is also very much related to the *sequential detection problem*, where a change in the distribution (of a series of random variables taken over time) may occur at some unknown time point, though the sample size is not prefixed. In Chapter 3, some quasi-sequential nonparametric procedures for the change-point problem are considered. This problem was not treated formally in the book, and some of the developments (particularly, the recursive procedures) have taken place only very recently. Basically, for various types of recursive nonparametric statistics, suitable invariance principles and consistency properties have been studied, and with the aid of these results, asymptotic theory analogous to the classical one in sequential analysis is developed. From the operational point of view, the testing procedures for the change-point model are quite similar to the time-sequential ones in Chapter 2, and the asymptotic theory run on parallel lines. Nonrecursive procedures (not properly belonging to the domain of sequential nonparametrics) have also been briefly referred to.

There are two fundamental problems in statistical estimation theory where fixed sample size solutions may not work out, and two (or multi-) stage or sequential procedures have distinct advantages; this feature is shared by both the parametric and nonparametric estimators. These are the *minimum risk* (*point*) *estimation* of a parameter (e.g., location) when the cost of sampling is incorporated in the *loss* function, and the *bounded-width confidence interval estimation* having a prescribed *coverage probability*. In Chapter 4, nonparametrics of these sequential estimation problems have been discussed. Procedures based on U-statistics (for general parameters) and robust (R-, L- and M-) estimators of location are considered. The generalized Behrens–Fisher model is also treated briefly (in a nonparametric setup). These theoretical developments demand some asymptotic representations for nonparametric estimators and certain uniform integrability results, which have interests of their own.

In medical trials, the patients may enter the clinic sequentially. This sequential recruitment of the units in a study is also quite common in many other problems in biostatistics and statistics. In such a case, instead of waiting until all the observations become available, one may perform interim analysis on the accumulating data, so that an early stopping may be achieved if the outcomes provocate so. In such *repeated significance testing* (on accumulating data), sufficient care must be taken in choosing the appropriate critical regions, such that the overall significance level is not out of control. Fortunately, in an asymptotic setup, the invariance principles for various nonparametric statistics can readily be incorporated in the formulation of repeated significance tests with prescribed overall significance levels. These are discussed in Chapter 5. In this context, some more recent developments on repeated significance *simultaneous testing* (and *multiple comparisons*) in a nonparametric setup are also reviewed.

Nonparametric analogues of the classical sequential probability (and likelihood) ratio tests are considered in Chapter 6. Here also, the basic invariance principles for some (usually aligned) nonparametric statistics and strong consistency results on some related estimators provide the access to the asymptotic theory of these tests, and they run quite parallel to the parametric cases. In fact, this asymptotic cohesiveness is utilized in the formulation of a measure of the *asymptotic relative efficiency* of competing tests in the sequential case (which extends the classical Pitman-efficiency result in the sequential case, without requiring the computations of the asymptotic ASN (average sample number) functions, and, thus, avoids some extra regularity conditions).

It may be remarked that a fairly detailed account of various invariance principles for various nonparametric statistics is given in the book referred to earlier. In view of this, in most of the cases in these lecture notes, the details of the proofs are omitted by suitable cross references. However, for a clear, motivating and smooth presentation of the theory, some of these mathematical presentations are necessary and retained. Further, the asymptotic theory has been presented here without any attempt to find out the degree of the approximations provided by the asymptotic theory when the sample sizes are only moderately large. From a theoretical point of view, these would require refinements on the existing results on the Berry–Esseen type bounds and Edgeworth type expansions for such sequential nonparametric statistics. Simply the order of the remainder terms may not be enough, and one may need to have precise bounds on these coefficients, so that their relevance to any sample size may be judged clearly. As of the present time, there remains much more to be accomplished in this direction before the theory can be implemented in practice. From the practical point of view, simulation studies can be undertaken to fathom the margin of errors of these asymptotic approximations. This approach has been fruitfully adapted in various specific problems in sequential nonparametrics, and the results are quite encouraging. Generally, these studies have revealed that the asymptotic results adapted to the finite sample case lead to some *conservativeness*. On the other hand, the adaptability of the asymptotic solutions in the nonparametric cases seems to be comparatively better than in their parametric counterparts. This robustness aspect deserves a lot of further investigation (both qualitatively and quantitatively). We propose this area for future research.

CHAPTER 2
Time-Sequential Nonparametrics

2.1. Introduction. In a *follow-up study*, arising in a *clinical trial* or a *life-testing problem*, typically, the *response* (viz. failure) occurs sequentially over time. Further, because of time, cost and other limitations, it may not always be possible to conduct the study until all the responses occur (and then to draw statistical conclusions). On the contrary, often a follow-up study is planned either for a fixed duration of time (*Type I censoring* or *truncation*) or for a period of time required to have a given number (or proportion) of responses (Type II censoring); in either scheme, the responses not occurring during the tenure of the study (i.e., the *censored observations*) induce some incompleteness, and thereby, introduce some complications in a valid and efficient statistical analysis of the experimental data. These Type I and Type II censoring schemes differ in the basic feature that for the former, the duration of the study is pre-fixed but the number of responses occurring during this period is random, while for the latter, this number is pre-fixed and the duration of the study is of stochastic nature. In most practical problems, unless the response distributions are, at least, roughly known, either of these single-point censoring schemes may lead to considerable loss of efficiency (relative to cost) of the statistical procedures. A too early termination may lead to an inadequate data set, and hence, may increase the risk of making incorrect statistical decisions, while unnecessary prolongation of the study may lead to consumption of valuable time (and cost), without any significant gain in the efficiency. For this reason, often, a *statistical monitoring* is advocated: The study is monitored from the beginning with the objective of an *early termination*, contingent on the statistical evidence from the accumulating data. This is referred to as a *progressive censoring scheme* (PCS). In clinical trials, a PCS is often advocated on the ground of *medical ethics* too. *Interim* (or *repeated*) *analysis* of accumulating clinical data is a common practice. For example, in a *treatment* vs. *control* study, instead of performing a terminal analysis at the projected endpoint of time, one may perform interim analysis at regular intervals of time (or at successive failure points), so that if at any intermediate point, a clear statistical difference is detected between the two sets of responses, then the study may be stopped and the surviving units be all switched to the better group. On the other hand, if no significant difference is perceived during the course of monitoring, then by continuing the study to the projected endpoint, no group is subjected to higher risk, and hence, conclusions based on the terminal (larger) data set should be more reliable. These interim analysis schemes may pose some technical problems, particularly from the statistical inferential point of view: Unless proper care is taken in the formulation of the inference procedure, repeated analysis of accumulating data may enhance the risk of making incorrect decisions. Further, in most follow-up studies, the accumulating data may not lead to stochastic processes with independent and/or stationary increments, and hence, the usual analysis schemes valid for such stationary processes may not be usable in such cases. Fortunately, the picture can be viewed as a *time-sequential scheme* for which a general class of nonparametric procedures works out nicely. In these developments, nonparametric procedures are adapted to PCS, some *martingale characterizations* are made for such PC nonparametric statistics, and this *martingale structure* paves the way for some *invariance principles* for these nonparametric statistics, and the theory of time-sequential nonparametrics rests heavily on these invariance principles. We shall discuss all these in the current chapter.

For the *simple regression model* (containing the *two-sample problem* as a particular case), a general class of *linear rank statistics* (LRS) was incorporated in the study of PCS and various properties were studied by Chatterjee and Sen (1973). Generalizations of these results to the *multiple regression model* (containing the *several-sample problem* as a special case) are due to Majumdar and Sen (1978a). LRS were also employed by Sen (1979b), (1981c) for the analysis of covariance (ANOCOVA) problem in a PCS. Our main interest will cluster around these developments.

The PC rank statistics are introduced in § 2.2, and the basic invariance principles are then presented in § 2.3. General properties of the nonparametric time-sequential procedures are considered in § 2.4. The ANOCOVA model in PCS is discussed in § 2.5. The problem of *staggered entry plans* and *dropouts* is considered in § 2.6. The last section deals with some other nonparametric and quasi-nonparametric procedures for PCS, with especial emphasis on the role of the weighted empirical processes and the impact of the *proportional hazard model* (Cox (1972)) in PCS.

2.2. Linear rank statistics under PCS. Let $X_1, \ldots, X_n$ be n independent random variables (r.v.) with continuous distribution functions (d.f.) $F_1, \ldots, F_n$, respectively, all defined on the real line $E(=(-\infty, \infty))$. Let R_{ni} be the *rank* of X_i among $X_1, \ldots, X_n$, for $i = 1, \ldots, n$ (ties neglected, with probability 1, by virtue of the assumed continuity of the F_i). Also, let $a_n(1), \ldots, a_n(n)$ a *set of scores* (real numbers), which we shall define more formally later on. Finally, let $\mathbf{c}_1, \ldots, \mathbf{c}_n$ be a set of q (≥ 1)-vectors of known constants. Then a (vector of) *linear rank statistics* (LRS), may be defined by

$$\mathbf{L}_n = \sum_{i=1}^{n} (\mathbf{c}_i - \bar{\mathbf{c}}_n) a_n(R_{ni}), \qquad (2.2.1)$$

where $\bar{\mathbf{c}}_n = n^{-1} \sum_{i=1}^{n} \mathbf{c}_i$. By suitable choice of the $\mathbf{c}_i$, these LRS encompass a broad range of rank statistics in current usage. For example, if $n = n_1 + n_2$, $n_1 \geq 1$, $n_2 \geq 1$ and $X_1, \ldots, X_{n_1}$ (and $X_{n_1+1}, \ldots, X_n$) are independent and identically distributed (i.i.d.) r.v. with a d.f. F (and G), then choosing $q = 1$ and $c_1 = \cdots = c_{n_1} = 0$, $c_{n_1+1} = \cdots = c_n = 1$, the LRS in (2.2.1) reduces to a conventional two-sample rank statistic: If we take $a_n(i) = i/(n+1)$, $1 \leq i \leq n$, L_n

reduces to the *Wilcoxon–Mann–Whitney* statistic; for $a_n(i) = -1 + \sum_{j=1}^{i}(n-j+1)^{-1}$, $1 \le i \le n$, to the *log-rank* statistic and for $a_n(i) =$ expected value of the ith order statistic of a sample of size n from the standard normal d.f., $1 \le i \le n$, it corresponds to the two-sample *normal scores* statistic. In the k (≥ 2) sample problem, we take $q = k - 1$, and the c_i can have then only k possible realizations $\mathbf{0}$, $(1, 0, \ldots, 0)'$, $(0, 1, 0, \ldots, 0)'$, $\ldots$, $(0, \ldots, 0, 1)$ (with respective frequencies $n_1, \ldots, n_k$). Then (2.2.1) reduces to the vector of multi-sample rank statistics. One may also consider the simple and multiple linear regression models, by letting

$$F_i(x) = F(x - \beta_0 - \boldsymbol{\beta}'\mathbf{c}_i), \qquad i = 1, \ldots, n, \quad x \in E, \tag{2.2.2}$$

where β_0 is an unknown parameter and $\boldsymbol{\beta}' = (\beta_1, \ldots, \beta_q)$ is the vector of regression coefficients. We may want to test then the null hypothesis of no regression (i.e., $\boldsymbol{\beta} = \mathbf{0}$), against appropriate alternatives.

In all the models considered above, the basic null hypothesis is the *hypothesis of randomness* (i.e., permutation invariance), viz.

$$H_0 : F_1 = \cdots = F_n = F \quad \text{(unknown)}. \tag{2.2.3}$$

However, in a follow-up study, the r.v. $X_1, \ldots, X_n$ are not observable at the beginning of the study, and hence, we need to attack the problem in a slightly different manner. Let $Z_{n,1} < \cdots < Z_{n,n}$ be the ordered r.v. corresponding to $X_1, \ldots, X_n$. Then, note that by definition (of the ranks), $X_i = Z_{n,R_{ni}}$ for $i = 1, \ldots, n$. We define the vector $\mathbf{S}_n = (S_{n1}, \ldots, S_{nn})$ of the *anti-ranks* S_{ni} by letting $Z_{n,i} = X_{S_{ni}}$, for $i = 1, \ldots, n$. Then, we have,

$$X_i = Z_{n,R_{ni}}, \quad Z_{n,i} = X_{S_{ni}}, \quad R_{nS_{ni}} = S_{nR_{ni}} = i, \quad 1 \le i \le n. \tag{2.2.4}$$

It is more convenient to describe the observable r.v. in a life testing setup in terms of the $Z_{n,i}$ and S_{ni}. Note that in a life testing problem, the ordered failures (i.e., $Z_{n,i}$) occur sequentially over time, so that at a time point t: $Z_{n,k} \le t < Z_{n,k+1}$, one has the set

$$(Z_{n,1}, S_{n1}), \ldots, (Z_{n,k}, S_{nk}) \tag{2.2.5}$$

of observable r.v., along with partial information that, for the remaining $n - k$ surviving units, the survival times are greater than t, for $k = 1, \ldots, n$, where, conventionally, we let $Z_{n,0} = -\infty$ and $Z_{n,n+1} = +\infty$. Thus, in a Type I censoring scheme, when the study is curtailed at a time point T ($<\infty$), we may define a nonnegative, integer valued r.v. $r(T) = \max\{k : Z_{n,k} \le T\}$, and then at the endpoint T, we have the set $\{(Z_{n,i}, S_{ni}), i \le r(T)\}$ of observable r.v.'s. We also have the additional information that the rest of the responses have magnitudes greater than T. In a Type II censoring scheme, for some pre-fixed r ($r \le n$), the study is stopped at the rth failure $Z_{n,r}$, so that the set of observable r.v.'s is $\{(Z_{n,i}, S_{ni}), i \le r\}$ and we also have the censoring information on the rest. Note that the length of the study time (i.e., $Z_{n,r}$) is a r.v. in this case. To adapt the

CHAPTER 2

LRS in either of these schemes, we make use of (2.2.1) and (2.2.4) and rewrite

$$\mathbf{L}_n = \sum_{i=1}^{n} (\mathbf{c}_{S_{ni}} - \bar{\mathbf{c}}_n) a_n(i). \tag{2.2.6}$$

Note that $\mathbf{L}_n$ does not depend on the $Z_{n,k}$ $(k \leq n)$ and is a sole function of the anti-ranks. As such, if we let $\mathbf{S}_n^{(r)} = (S_{n1}, \dots, S_{nr})$ and denote by $\mathscr{B}_{nr}$, the sigma-field generated by $\mathbf{S}_n^{(r)}$, for $r = 1, \dots, n$ (and let $\mathscr{B}_{n0}$ be the trivial sigma-field), then $\mathscr{B}_{nr}$ is nondecreasing in r $(0 \leq r \leq n)$. Then, the projection

$$E_0\{\mathbf{L}_n \mid \mathscr{B}_{nr}\} = \mathbf{L}_{nr} \quad \text{(say)} \tag{2.2.7}$$

(where E_0 denotes the (conditional) expectation under H_0) is termed a censored LRS, where the censoring is made at the rth failure $Z_{n,r}$. Note that under H_0, $\mathbf{S}_n^{(r)}$ takes on each realization $(i_1, \dots, i_r)$ (where $1 \leq i_1 \neq \cdots \neq i_r \leq n$) with the same probability $\{n \cdots (n-r+1)\}^{-1}$, while given $\mathbf{S}_n^{(r)}$, S_{ni} (for any $i : r+1 \leq i \leq n$) can assume any value of the complementary set $\{1, \dots, n\} \backslash \mathbf{S}_n^{(r)}$ with the same probability $(n-r)^{-1}$. Hence, (2.2.7) reduces to

$$\mathbf{L}_{nr} = \sum_{i=1}^{r} (\mathbf{c}_{S_{ni}} - \bar{\mathbf{c}}_n)[a_n(i) - a_n^*(r)], \tag{2.2.8}$$

where $a_n^*(n) = 0$ (conventionally) and for $r \leq n-1$,

$$a_n^*(r) = (n-r)^{-1} \sum_{j=r+1}^{n} a_n(j). \tag{2.2.9}$$

Note that for any given (n, r), under H_0, by the uniform distribution of $\mathbf{S}_n^{(r)}$, we have

$$E_0 \mathbf{L}_{nr} = \mathbf{0} \quad \text{and} \quad E_0 \mathbf{L}_{nr} \mathbf{L}'_{nr} = A_{n,r}^2 \cdot \mathbf{C}_n \tag{2.2.10}$$

where

$$\mathbf{C}_n = \sum_{i=1}^{n} (\mathbf{c}_i - \bar{\mathbf{c}}_n)(\mathbf{c}_i - \bar{\mathbf{c}}_n)', \tag{2.2.11}$$

$$\bar{a}_n = n^{-1} \sum_{i=1}^{n} a_n(i), \tag{2.2.12}$$

$$A_{n,r}^2 = \frac{1}{n-1} \left\{ \sum_{i=1}^{r} a_n^2(i) + (n-r)[a_n^*(r)]^2 - n\bar{a}_n^2 \right\}. \tag{2.2.13}$$

Thus, whenever $\mathbf{C}_n$ is positive definite (p.d.) and $A_{n,r}^2 > 0$, one may construct a test statistic

$$\mathscr{L}_{nr} = A_{n,r}^{-2}(\mathbf{L}'_{nr} \mathbf{C}_n^{-1} \mathbf{L}_{nr}), \tag{2.2.14}$$

and under H_0, for every given (n, r), $\mathscr{L}_{nr}$ is a distribution-free statistic. This explains the distribution-free structure of rank statistics under Type II censoring. Under Type I censoring, $r(T)$ is itself a r.v., so that $\mathscr{L}_{nr(T)}$ may have a distribution dependent on F through the d.f. of $r(T)$. Note that under H_0,

$$P\{r(T) = r\} = \binom{n}{r}[F(T)]^r[1 - F(T)]^{n-r}, \qquad 0 \leq r \leq n, \qquad (2.2.15)$$

so that under H_0, $\mathscr{L}_{nr(T)}$ may not be genuinely distribution-free. However, under H_0, the $Z_{n,i}$ and $\mathbf{S}_n$ are mutually independent, while $r(T)$ depends only on the $Z_{n,i}$. Hence, given $r(T) = r$, $\mathscr{L}_{nr(T)}$ has the same distribution as of $\mathscr{L}_{nr}$, when H_0 holds. Hence, under Type I censoring, $\mathscr{L}_{nr(T)}$ is a conditionally distribution-free statistic. (For $q = 1$, one may also use one-sided versions of Type I and Type II censored tests based on the L_{nr}.)

As has been mentioned earlier, in a PCS one usually monitors the study from the beginning with a view to have a possible early stopping. In this setup, at the kth failure $Z_{n,k}$ ($k \geq 1$), one has the set of observable r.v.'s in (2.2.6), so that one may compute $\mathscr{L}_{nk}$ as in (2.2.14). Thus, one encounters the sequence of test statistics

$$\{\mathscr{L}_{nk}, 0 \leq k \leq n\} \quad \text{(where } \mathscr{L}_{n0} = 0); \qquad (2.2.16)$$

and based on this triangular array of statistics, the basic problem is to define a *stopping rule* by which the time-sequential procedure is well defined, and then to study the desirable properties of the stopping rule and the test procedure. In the classical sequential analysis, one usually encounters a stochastic process with independent and homogeneous increments. However, in the given context, for the $\mathbf{L}_{nr}$ (or the corresponding $\mathscr{L}_{nr}$), we do not have (generally) independent or homogeneous increments (even under H_0). In this context, some *invariance principles* (relating to the $\mathbf{L}_{nr}$ or the $\mathscr{L}_{nr}$) based on certain *martingale characterizations of* PC LRS give us the desired access for the development of the theory of time-sequential tests, parallel to the classical sequential tests, and these will be discussed first in the next section.

2.3. Invariance principles for PC LRS. Note that by definition in (2.2.7), *under* H_0, *for every* n (≥ 1), $\{\mathbf{L}_{nk}, \mathscr{B}_{nk}; 0 \leq k \leq n\}$ *is a martingale* (vector) *sequence, closed on the right by* $\mathbf{L}_n$. Further, by direct computations,

$$A_{n,r+1}^2 - A_{n,r}^2 = (n-1)^{-1}(n-r)^{-1}(n-r-1)[a_n(r+1) - a_n^*(r+1)]^2 \geq 0, \qquad (2.3.1)$$

for every $r: 0 \leq r \leq n-1$, so that for every n (≥ 1),

$$A_{n,r}^2 \text{ is nondecreasing in } r: 0 \leq r \leq n; \qquad (2.3.2)$$

the last property also follows from (2.2.10) and the martingale property. We may also note that on letting $b_{nr}(i) = a_n(i)$, $i \leq r$ and $a_n^*(r)$, $r+1 \leq i \leq n$, we may write $\mathbf{L}_{nr}$ in (2.2.8) as

$$\sum_{i=1}^{n} (\bar{\mathbf{c}}_i - \bar{\mathbf{c}}_n) b_{nr}(R_{ni}). \qquad (2.3.3)$$

This latter representation enables one to use the permutational central limit theorem (PCLT) (in its most general form, due to Hájek (1961)) and to conclude that under H_0 and appropriate regularity conditions, as $n \to \infty$,

$$A_{n,r}^{-1}\mathbf{C}_n^{-1/2}\mathbf{L}_{nr} \sim \mathcal{N}_q(\mathbf{0}, \mathbf{I}_q) \tag{2.3.4}$$

where $\mathbf{D}_n = \mathbf{C}_n^{-1/2}$ is defined by $\mathbf{D}_n \mathbf{C}_n \mathbf{D}_n' = \mathbf{I}_q$, the identity matrix of order q, and tacitly, $\mathbf{C}_n$ is assumed to be p.d. We intend to consider a weak invariance principle which extends (2.3.4) to the (partial) sequence in (2.2.16).

For this purpose, we define the *scores* formally as

$$a_n(i) = E\phi(U_{ni}) \quad \text{or} \quad \phi(i/(n+1)), \quad 1 \le i \le n, \tag{2.3.5}$$

where $U_{n1} < \cdots < U_{nn}$ are the ordered r.v.'s of a sample of size n from the uniform $(0, 1)$ d.f., and the *score function* $\phi = \{\phi(t), 0 < t < 1\}$ is assumed to be expressible as the difference of two nondecreasing and square integrable functions (ϕ_1 and ϕ_2) inside $(0, 1)$. Further, we assume that $\mathbf{C}_n$ is p.d. for every $n \ge n_0 (\ge q)$ and the following (generalized) Noether condition holds:

$$\max\{(\mathbf{c}_k - \bar{\mathbf{c}}_n)'\mathbf{C}_n^{-1}(\mathbf{c}_k - \bar{\mathbf{c}}_n) : k \in [1, n]\} \to 0 \quad \text{as } n \to \infty. \tag{2.3.6}$$

Now, for a given pair (r, n) $(1 \le r \le n)$, let us define

$$k_{nr}(t) = \max\{k : A_{n,k}^2 \le tA_{n,r}^2\}, \quad 0 \le t \le 1, \tag{2.3.7}$$

so that $k_{nr}(t)$ is a nondecreasing, right-continuous and nonnegative integer valued function. Then, for every (n, r), we define (vector-valued) stochastic processes $\mathbf{W}_{nr}^{(j)} = \{\mathbf{W}_{nr}^{(j)}(t), 0 \le t \le 1\}$, $j = 1, 2$, by letting

$$\mathbf{W}_{nr}^{(1)}(t) = A_{n,r}^{-1}\mathbf{C}_n^{-1/2}\mathbf{L}_{nk_{nr}(t)}, \quad 0 \le t \le 1, \tag{2.3.8}$$

$$\mathbf{W}_{nr}^{(2)}(t) = A_{n,k_{nr}(t)}^{-1}\mathbf{C}_n^{-1/2}\mathbf{L}_{nk_{nr}(t)}, \quad 0 \le t \le 1 \tag{2.3.9}$$

where, in the second case, the lower endpoint 0 is excluded as $A_{n,0} = 0$. Side by side, we introduce the processes $B_{nr}^{(j)} = \{B_{nr}^{(j)}(t), 0 \le t \le 1\}$ by letting

$$[B_{nr}^{(1)}(t)]^2 = [\mathbf{W}_{nr}^{(1)}(t)]'[\mathbf{W}_{nr}^{(1)}(t)]$$
$$= A_{n,r}^{-2}(\mathbf{L}_{nk_{nr}(t)}'\mathbf{C}_n^{-1}\mathbf{L}_{nk_{nr}(t)}), \quad 0 \le t \le 1, \tag{2.3.10}$$

$$[B_{nr}^{(2)}(t)]^2 = [\mathbf{W}_{nr}^{(2)}(t)]'[\mathbf{W}_{nr}^{(2)}(t)]$$
$$= A_{n,k_{nr}(t)}^{-2}(\mathbf{L}_{nk_{nr}(t)}'\mathbf{C}_n^{-1}\mathbf{L}_{nk_{nr}(t)}), \quad 0 < t \le 1. \tag{2.3.11}$$

Let now $\xi_j = \{\xi_t(t), 0 \le t \le 1\}$, $j = 1, \dots, q$ be q independent copies of a standard Wiener process on $[0, 1]$ and let

$$\boldsymbol{\xi}^{(1)} = \{(\xi_1(t), \dots, \xi_q(t))', 0 \le t \le 1\}, \tag{2.3.12}$$

$$\boldsymbol{\xi}^{(2)} = \{t^{-1/2}(\xi_1(t), \dots, \xi_q(t))', 0 < t \le 1\}, \tag{2.3.13}$$

$$B_q = \{B_q(t) = [\{\boldsymbol{\xi}^{(1)}(t)\}'\{\boldsymbol{\xi}^{(1)}(t)\}]^{1/2}, 0 \le t \le 1\}, \tag{2.3.14}$$

$$B_q^* = \{B_q^*(t) = [\{\boldsymbol{\xi}^{(2)}(t)\}'\{\boldsymbol{\xi}^{(2)}(t)\}]^{1/2}, 0 < t \le 1\}. \tag{2.3.15}$$

Thus, B_q is a (q-parameter) *Bessel process* and B_q^* is the *standardized Bessel* *process.* For every $\varepsilon : 0 < \varepsilon < 1$, $_\varepsilon Y$ stands for the process $\{Y(t), \varepsilon \leq t \leq 1\}$. Then, we have the following

THEOREM 2.3.1. *Under H_0 in (2.2.3) and the regularity conditions in (2.3.5)–(2.3.6), whenever $r/n \to p$: $0 < p \leq 1$,*

$$\mathbf{W}_{nr}^{(1)} \underset{\mathscr{D}}{\to} \boldsymbol{\xi}^{(1)} \quad \text{in the } J_1\text{-topology on } D^q[0, 1], \tag{2.3.16}$$

$$B_{nr}^{(1)} \underset{\mathscr{D}}{\to} B_q \quad \text{in the } J_1\text{-topology on } D[0, 1], \tag{2.3.17}$$

and for every ε $(0 < \varepsilon < 1)$,

$$_\varepsilon \mathbf{W}_{nr}^{(2)} \underset{\mathscr{D}}{\to} {}_\varepsilon \boldsymbol{\xi}^{(2)} \quad \text{in the } J_1\text{-topology on } D^q[\varepsilon, 1], \tag{2.3.18}$$

$$_\varepsilon B_{nr}^{(2)} \underset{\mathscr{D}}{\to} {}_\varepsilon B_q^* \quad \text{in the } J_1\text{-topology on } D[\varepsilon, 1]. \tag{2.3.19}$$

(For $q = 1$, the results are due to Chatterjee and Sen (1973), while the general case of $q \geq 1$ is treated in Majumdar and Sen (1978a).) By virtue of the martingale structure of the $\{L_{n,k} : 0 \leq k \leq n\}$, $n \geq 1$ (under H_0), the proof of the theorem rests directly on the weak convergence of martingale processes (in the Skorokhod J_1-topology and "sup-norm" metric); this is discussed in detail in Sen (1981d, Chap. 11).

Let us now consider the case where H_0 in (2.2.3) may not hold. It is clear from (2.2.7)–(2.2.8) that when H_0 does not hold, the martingale structure may not hold also. For arbitrary $F_1, \ldots, F_n$, one may appeal to the general results of Hájek (1968) and, using the Cramér–Wold characterization of multi-normality, show that, for an arbitrary m (≥ 1) and $r_1, \ldots, r_m$ (such that $n^{-1} r_j \to \pi_j$; $0 \leq \pi_1 < \cdots < \pi_m \leq 1$), the joint d.f. of $\mathbf{C}_n^{-1/2}(\mathbf{L}_{nr_1} - \boldsymbol{\mu}_{nr_1}, \ldots, \mathbf{L}_{nr_m} - \boldsymbol{\mu}_{nr_m})$, for suitable (nonstochastic) normalizing vectors $\boldsymbol{\mu}_{nr_1}, \ldots, \boldsymbol{\mu}_{nr_m}$, is asymptotically multi-normal. However, the covariance structure of this asymptotic multi-normal distribution, in general, may not agree with the covariance structure of $\boldsymbol{\xi}^{(1)}$, so that even if the *tightness* of $W_{nr}^{(1)}$ (or $W_{nr}^{(2)}$) may be established (under some additional regularity conditions), the weak convergence of $W_{nr}^{(1)}$ (or $W_{nr}^{(2)}$) to $\boldsymbol{\xi}^{(1)}$ (or $\boldsymbol{\xi}^{(2)}$), with an appropriate drift function, may not be generally true. For the verification of the above tightness condition, in the absence of a martingale (or sub-martingale) structure, one may have to verify some "Billingsley-type" inequalities (viz. Billingsley (1968, pp. 98–102)), and this may demand more restrictive conditions on the score function as well as on the d.f. $F_1, \ldots, F_n$. Actually, for arbitrary $F_1, \ldots, F_n$, weak convergence to a drifted $\boldsymbol{\xi}^{(1)}$ (or $\boldsymbol{\xi}^{(2)}$), in general, does not hold. From the point of view of asymptotic power properties of nonparametric time-sequential tests, one may like to confine oneself to some "local" alternatives, for which the asymptotic power does not converge to its upper asymptote 1. This is usually done by assuming that

$$\max_{1 \leq i \neq j \leq n} \sup_x |F_i(x) - F_j(x)| = O(n^{-1/2}), \tag{2.3.20}$$

$$n^{-1} \mathbf{C}_n \to \mathbf{C}^0 \quad \text{(p.d.)} \quad \text{as } n \to \infty \tag{2.3.21}$$

In this context, one can introduce some regularity conditions on the F_i (typically, requiring them to have absolutely continuous density functions with finite *Fisher information*) and show that such a sequence of local alternatives satisfies the condition of *contiguity* with respect to the null case (see Hájek and Šidák (1967, Chap. 6) for some very novel discussions). For such *contiguous alternatives*, Theorem 2.3.1 extends in a very natural way. First, the tightness of the $W_{nr}^{(j)}$ under H_0 (in Theorem 2.3.1) and the contiguity of the alternatives ensure the tightness under such alternatives too (see Sen (1981d, Thm. 4.3.4) in this context). Secondly, under such contiguous alternatives, the covariance structure of the asymptotic distribution of the $W_{nr}^{(j)}(t_k)$ agrees with the null hypothesis case, while the normalizing vectors $\boldsymbol{\mu}_{nr}(t)$ have nice asymptotic expressions (though not, in general, linear in t). Thus, here, one can obtain weak convergence to appropriate drifted $\boldsymbol{\xi}^{(j)}$. For details, we may refer to Majumdar and Sen (1978a), DeLong (1980) and Sen (1981d, Chap. 11).

2.4. Time-sequential tests: General properties. We would like to incorporate the set of statistics in (2.2.16) in the formulation of suitable time-sequential tests based on some well defined stopping rules. In this context, we may adopt the following two schemes (among other possibilities, of course):

(a) *Analogues of* CUSUM *procedures.* We may view the $\mathbf{L}_{nk}$ in (2.2.8) as analogues of the usual cumulative sums (though the underlying setups are not the same). In that case, if one decides to continue the study up to a maximum period and needs to have r failures, for some $r: r/n \to \theta: 0 < \theta \le 1$, then one may consider the test statistic as

$$\max \{ \mathbf{L}'_{nk} \mathbf{C}_n^{-1} \mathbf{L}_{nk} : k \le r \}, \tag{2.4.1}$$

so that by (2.2.14) and (2.4.1), we have this equivalent to

$$\max \{ (A_{n,k}^2 / A_{n,r}^2) \mathcal{L}_{nk} : k \le r \}. \tag{2.4.2}$$

Coupled with (2.3.8) and (2.3.10), (2.4.2) reduces to

$$[\sup \{ B_{nr}^{(1)}(t) : 0 \le t \le 1 \}]^2. \tag{2.4.3}$$

In this case, the *stopping rule* $K_{nr}^{(1)}$ is defined by

$$K_{nr}^{(1)} = \begin{cases} \min \{ k : k \le r \text{ and } \mathbf{L}'_{nk} \mathbf{C}_n^{-1} \mathbf{L}_{nk} \ge v_{nr}^{(\alpha)} \}, \\ r \quad \text{if } \mathbf{L}'_{nk} \mathbf{C}_n^{-1} \mathbf{L}_{nk} < v_{nr}^{(\alpha)} \quad \forall k \le r, \end{cases} \tag{2.4.4}$$

where $v_{nr}^{(\alpha)}$ is a constant (depending on the *significance level* α $(0 < \alpha < 1)$ and (n, r)) such that

$$P \left\{ \max_{k \le r} \mathbf{L}'_{nk} \mathbf{C}_n^{-1} \mathbf{L}_{nk} \ge v_{nr}^{(\alpha)} \mid H_0 \right\} \le \alpha. \tag{2.4.5}$$

Note that K_{nr} is bounded from above by r, with probability 1, and in this setup, one may even take $r = n$, though generally r is taken smaller than n (i.e., $r/n \to \theta : \theta < 1$). Further, if $\delta_{nr}^{(\alpha)}$ be the upper $100\alpha\%$ point of the distribution of $\sup \{ B_{nr}^{(1)}(t) : 0 \le t \le 1 \}$, then $v_{nr}^{(\alpha)} = A_{n,r}^2 (\delta_{nr}^{(\alpha)})^2$.

For small values of (n, r), the distribution of the statistic in (2.4.3) can be obtained by direct enumeration of the $\{n \cdots (n-r+1)\}$ possible realizations of $\mathbf{S}_n^{(r)}$, and hence, $\delta_{nr}^{(\alpha)}$ can be obtained. This process becomes prohibitively laborious as n increases. By virtue of Theorem 2.3.1, we may, however, provide the following large sample approximation. Note that if $r/n \rightarrow \theta : 0 < \theta < 1$, then

$$A_{n,r}^2 \rightarrow A_\theta^2 = \int_0^\theta \phi^2(u)\, du + (1-\theta)^{-1} \left(\int_\theta^1 \phi(u)\, du \right)^2 - \left(\int_0^1 \phi(u)\, du \right)^2.$$

$$(2.4.6)$$

Further, if $\delta_q^{(\alpha)}$ is the upper $100\alpha\%$ point of the distribution of $\sup \{B_q(t) : 0 \le t \le 1\}$, then by (2.3.17), $\delta_{nr}^{(\alpha)} \rightarrow \delta_q^{(\alpha)}$ as $n \rightarrow \infty$. Consequently, we obtain that as n increases,

$$\nu_{nr}^{(\alpha)} \rightarrow A_\theta^2 (\delta_q^{(\alpha)})^2 = \nu_\theta^{(\alpha)} \quad \text{(say)}.$$

$$(2.4.6)$$

For values of $\delta_q^{(\alpha)}$, for various $q\ (\ge 1)$ and α, we may refer to DeLong (1980). For $q = 1$, we may note that for every $\lambda \ge 0$,

$$P\left\{ \sup_{0 \le t \le 1} \xi_1(t) > \lambda \right\} = 2[1 - \Phi(\lambda)],$$

$$(2.4.7)$$

$$P\left\{ \sup_{0 \le t \le 1} |\xi_1(t)| > \lambda \right\} = 1 - \sum_{k=-\infty}^\infty (-1)^k \{\Phi((2k+1)\lambda) - \Phi((2k-1)\lambda)\},$$

$$(2.4.8)$$

where Φ is the standard normal d.f., and hence, the critical values may also be obtained from the standard normal distributional tables.

Operationally, the time-sequential procedure consists in computing at each failure point $(Z_{n,k})$ the rank test statistic $\mathscr{L}_{nk}$, in (2.2.8). So long as $\mathscr{L}_{nk}$ lies below $\nu_{nr}^{(\alpha)}/A_{n,k}^2$, the process continues. If, either for some $k\ (\le r)$, $\mathscr{L}_{nk}$, for the first time, exceeds $\nu_{nr}^{(\alpha)}/A_{n,k}^2$, the study is curtailed at that point along with the rejection of H_0. If no such k exists, then at the preplanned rth failure point $Z_{n,r}$, the study is completed, along with the acceptance of H_0.

From Theorem 2.3.1 and (2.4.6), we may conclude that the level of significance of this time-sequential test is asymptotically equal to $\alpha\ (0 < \alpha < 1)$. Further, note that by (2.4.4), for every $k < r$

$$P\{K_{nr}^{(1)} > k\} = P\left\{ \max_{1 \le j \le k} \mathbf{L}_{nj}' \mathbf{C}_n^{-1} \mathbf{L}_{nj} < \nu_{nr}^{(k)} \right\}$$

$$= P\left\{ \sup_{0 \le t \le 1} [B_{nk}^{(1)}(t)]^2 < \nu_{nr}^{(k)}/A_{n,k}^2 \right\}.$$

$$(2.4.9)$$

Thus, if we let $k = ur$, $0 < u \le 1$, then by (2.4.6) and (2.4.9), for every $u : 0 < u < 1$, under H_0,

$$P\{r^{-1} K_{nr}^{(1)} > u\} \rightarrow P\left\{ \sup_{0 \le t \le 1} B_q(t) < \delta_q^{(\alpha)} A_\theta/A_{u\theta} \right\},$$

$$(2.4.10)$$

where $r/n \to \theta : 0 < \theta \leqq 1$. We may also note that

$$E_0\{r^{-1}K_{nr}^{(1)}\} \to \int_0^1 P\{r^{-1}K_{nr}^{(1)} > u\} \, du$$

$$\to \int_0^1 P\left\{\sup_{0 \leqq t \leqq 1} B_q(t) < \delta_q^{(\alpha)} A_\theta / A_{u\theta}\right\} du, \qquad (2.4.11)$$

where in this moment convergence result, we have made use of the fact that $r^{-1}K_{nr}^{(1)}$ is bounded from above by one (with probability 1), so that (2.4.10) ensures (2.4.11). Note that for every $u : 0 < u < 1$, $\delta_q^{(\alpha)} A_\theta / A_{u\theta} > \delta_q^{(\alpha)}$, so that the right-hand side of (2.4.10) is bounded from below by $1 - \alpha$, and hence, the right-hand side of (2.4.11) is also bounded from below by $1 - \alpha$. Or, in other words, under H_0, we do not have a small expected stopping time, and this result is quite intuitive. Also, note that one may rewrite the right-hand side of (2.4.11) as

$$\int_0^1 P\left\{\sup_{0 \leqq t \leqq A_{u\theta}^2 / A_\theta^2} B_q(t) < \delta_q^{(\alpha)}\right\} du, \qquad (2.4.12)$$

and the same conclusion follows from the monotonicity of the $\sup\{B_q(t), 0 \leqq t \leqq s\}$ in $s \in [0, 1]$.

Consider next the nonnull case (contiguous alternatives $\{H_n\}$) under which the asymptotic results are tractable and have meaningful interpretations. Here, if we consider the asymptotic distribution of $\mathbf{W}_{nr}^{(1)}(t)$, the corresponding normalizing function $\boldsymbol{\mu}_{nr}(t)$, when incorporated in the definition of the $B_{nr}^{(1)}(t)$ in (2.3.10), leads to a drift function, which we denote by $\beta(t; \phi)$, $0 \leqq t \leqq 1$. Thus, under $\{H_n\}$,

$$B_{nr}^{(1)} \xrightarrow{\mathcal{D}} \{B_q(t) + \beta(t; \phi), 0 \leqq t \leqq 1\} \quad \text{as } n \to \infty. \qquad (2.4.13)$$

Hence, the *asymptotic power function* of the test, under $\{H_n\}$, may be expressed as

$$P\{B_q(t) + \beta(t; \phi) > \delta_q^{(\alpha)}, \text{ for some } t : 0 \leqq t \leqq 1\}. \qquad (2.4.14)$$

If there exists an (optimal) score function ϕ^0 such that

$$\beta(t; \phi^0) \geqq \beta(t; \phi) \quad \forall 0 \leqq t \leqq 1, \quad \text{all } \phi, \qquad (2.4.15)$$

(with at least a strict inequality for some t), then by (2.4.14)–(2.4.15), for the time-sequential test based on this score function ϕ^0, the asymptotic power function is maximized (under $\{H_n\}$). Such asymptotically optimal score functions were studied (for $q = 1$) by Sen (1976b) and reported in Sen (1981d, Chap. 11); a very similar picture holds for $q \geqq 1$. Also, note that by (2.4.9) and (2.4.13), the asymptotic expression for the expected value of $r^{-1}K_{nr}^{(1)}$, under H_n, is given by

$$\int_0^1 P\left\{\sup_{0 \leqq t \leqq A_{u\theta}^2 / A_\theta^2} \{B_q(t) + \beta(t, \phi)\} < \delta_q^{(\alpha)}\right\} du. \qquad (2.4.16)$$

Such a result has been used by DeLong (1980) in comparing the expected stopping times under local alternatives. Here also, (2.4.15) leads to a minimization of (2.4.16), so that whenever such asymptotically optimal score functions exist, they lead to asymptotically minimum expected stopping times too.

(b) *Analogues of RST procedures.* We may view the PCS in the light of repeated significance tests (RST), where, at each failure, one is essentially making a test for H_0 based on the censored rank statistic $\mathcal{L}_{nk}$, $k \leq r$. Thus, if one decides to conduct the study up to a maximum period during which r failures would occur, then an appropriate test statistic is

$$\max \{\mathcal{L}_{nk} : k \leq r\}, \tag{2.4.17}$$

In this case, the *stopping rule* $K_{nr}^{(2)}$ is defined by

$$K_{nr}^{(2)} = \begin{cases} \min \{k : \mathcal{L}_{nk} \geq \mathcal{L}_{n(\alpha)}^*, k \leq r\}, \\ r \quad \text{if } \mathcal{L}_{nk} < \mathcal{L}_{n(\alpha)}^*, \ \forall k \leq r, \end{cases} \tag{2.4.18}$$

where $\mathcal{L}_{n(\alpha)}^*$ is so chosen that

$$P\{\mathcal{L}_{nk} \geq \mathcal{L}_{n(\alpha)}^* \text{ for some } k \leq r \mid H_0\} \leq \alpha, \tag{2.4.19}$$

and α $(0 < \alpha < 1)$ is the significance level of the test.

Though in small samples, $\mathcal{L}_{n(\alpha)}^*$ may be determined by enumeration of the $\{n \cdots (n - r + 1)\}$ equally likely realizations of $\mathbf{S}_n^{(r)}$ (and the corresponding values of $\mathcal{L}_{nk}$, $k \leq r$), the process not only becomes prohibitively laborious for large n, but also encounters a small technical problem. As n becomes large, for small values of k, the $\mathcal{L}_{nk}$ may fluctuate quite irregularly, and thus, may distort (2.4.9) for any approximating $\mathcal{L}_{n(\alpha)}^*$. This difficulty can easily be overcome by choosing an ε $(0 < \varepsilon < 1)$ and a k_n such that

$$A_{n,k_n}^2 / A_{n,r}^2 \geq \varepsilon > 0. \tag{2.4.20}$$

In that case, the repeated significance testing starts only at the k_nth failure and then proceeds on as needed. Thus, in (2.4.18)–(2.4.19), one needs to restrict k to $k_n \leq k \leq r$. Let then $_\varepsilon \delta_{q*}^{(\alpha)}$ be the upper $100\alpha \%$ point of the distribution of $\sup [B_q^*(t) : \varepsilon \leq t \leq 1]$; for various q (≥ 1) and ε $(0 < \varepsilon < 1)$ and α $(0 < \alpha < 1)$, these critical values have been obtained by DeLong (1981). Then, by Theorem 2.3.1, we have

$$\mathcal{L}_{n(\alpha)}^* \to {_\varepsilon \delta_{q*}^{(\alpha)}} \quad \text{as } n \to \infty \quad (\forall \, 0 < \varepsilon < 1, 0 < \alpha < 1). \tag{2.4.21}$$

Operationally, the testing procedure consists in computing the $\mathcal{L}_{nk}$ at each failure point $Z_{n,k}$, for $k \geq k_n$. If, for the first time, for some k $(\leq r)$, $\mathcal{L}_{nk}$ exceeds $\mathcal{L}_{n(\alpha)}^*$, the study is curtailed along with the rejection of H_0. If no such k exists, the study is concluded at the preplanned rth failure point $(Z_{n,r})$ and the null hypothesis is accepted.

From Theorem 2.3.1 and (2.4.21), we conclude that the level of significance of the test is (asymptotically) equal to α. Further, by (2.4.18), for every

$k : k_n \leqq k < r,$

$$P\{K_{nr}^{(2)} > k\} = P\left\{ \max_{k_n \leqq j \leqq k} \mathscr{L}_{nk} < \mathscr{L}_{n(\alpha)}^* \right\}$$

$$= P\left\{ \sup_{\varepsilon \leqq t \leqq A_{n,k}^2/A_{n,r}^2} B_{nr}^{(2)}(t) < \mathscr{L}_{n(\alpha)}^* \right\}. \qquad (2.4.22)$$

Thus, if we let $k = ru$ $(0 < u < 1)$ and $r/n \to \theta : 0 < \theta \leqq 1$, then

$$E_0\{r^{-1}K_{nr}^{(2)}\} \to \int_\varepsilon^1 P\left\{ \sup_{\varepsilon \leqq t \leqq A_{u\theta}^w/A_\theta^2} B_q^*(t) < {}_\varepsilon\delta_{q*}^{(\alpha)} \right\} du. \qquad (2.4.23)$$

Again, by virtue of the fact that $A_{u\theta}^2/A_\theta^2$ is $\leqq 1$, $\forall u \in (0, 1)$, we conclude that the right-hand side of (2.4.23) is $\geqq 1 - \alpha$, so that the expected stopping time (under H_0) is not very small (as it should be).

Here also, we may consider the case of contiguous alternatives ($\{H_n\}$), under which the asymptotic power function is

$$P\{t^{-1/2}(b_q(t) + \beta(t; \phi)) > {}_\varepsilon\delta_{q*}^{(\alpha)}, \text{ for some } t : \varepsilon \leqq t \leqq 1\}, \qquad (2.4.24)$$

while the asymptotic form of the expected stopping number (i.e., $E(r^{-1}K_{nr}^{(2)} \mid H_n)]$ is given by

$$\varepsilon + \int_\varepsilon^1 P\{t^{-1/2}(B_q(t) + \beta(t; \phi)) > {}_\varepsilon\delta_{q*}^{(\alpha)}, \text{ for some } \varepsilon \leqq t \leqq u\} du. \qquad (2.4.25)$$

Again, (2.4.15) provides a sufficient condition for the asymptotic optimality in terms of maximum asymptotic power or minimum expected stopping number.

In passing, we may remark that the choice of ε in this procedure is quite critical. For very small values of ε, $\mathscr{L}_{n(\alpha)}^*$ may become quite large, and may pull down the asymptotic power, while for ε away from 0, one essentially ignores the early part of the study, which could have, otherwise, led to the rejection of H_0, when it is not true. As a compromise, $\varepsilon = 0.05$ or 0.1 looks quite suitable.

The two procedures essentially relate to the Bessel process with horizontal and square root (time) boundaries, respectively. Thus, there may be certain alternatives for which one test will have better asymptotic power properties (than the other), and, it may not be possible to say, categorically, that one is a better procedure for all alternatives. However, the simulation studies made on the asymptotic power properties of the two procedures (in some typical situations) show that the performances are quite similar to each other.

2.5. Tests for ANOCOVA under PCS. Nonparametric tests for ANOCOVA are available in the literature, and, at least asymptotically, they perform better than the corresponding tests for ANOVA, where the concomitant variates are ignored. In the context of follow-up studies too, time-sequential tests incorporating the information in the concomitant variates are available in the literature and will be discussed in this section.

Besides the primary variate X_0, we assume that there is a vector $\mathbf{X}^*$ of covariates, which are observable at the starting point of the study; the primary variate is subject to the typical life testing setup, discussed in § 2.1. Thus, we have a set of independent vectors $\mathbf{X}_i = (X_{0i}, \mathbf{X}_i^{*\prime})'$, $i = 1, \ldots, n$. We denote the marginal (joint) d.f. of $\mathbf{X}_i^*$ by F_i^*, and, in the conventional way, we assume that

$$F_1^* = \cdots = F_n^* = F^* \quad \text{(unknown).} \tag{2.5.1}$$

Further, we denote the conditional d.f. of X_{0i}, given $\mathbf{X}_i^* = \mathbf{x}^*$, by $F_i^0(x \mid \mathbf{x}^*)$, $i = 1, \ldots, n$. Our null hypothesis of interest is

$$H_0 : F_1^0(x \mid \mathbf{x}^*) = \cdots = F_n^0(x \mid \mathbf{x}^*) = F^0(x \mid \mathbf{x}^*) \quad \text{(unknown).} \tag{2.5.2}$$

In view of the fact that the responses X_{0i} occur sequentially over time, we intend to consider some time-sequential nonparametric tests which are the natural generalizations of the ones in § 2.4.

We write $\mathbf{X}_i^* = (x_{xi}^*, \ldots, X_{pi}^*)'$, so that the $\mathbf{X}_i$ are $(p+1)$-vectors, for some $p \geq 1$. Consider the $(p+1) \times n$ matrix $(\mathbf{X}_1, \ldots, \mathbf{X}_n)$ and within each row rank the elements—the corresponding rank matrix is denoted by $\mathbf{R}_n$ and the elements in the jth row are denoted by R_{ji}, $1 \leq i \leq n$, for $j = 0, 1, \ldots, p$. To be more flexible, for the jth row, we take the scores as $a_{nj}(i)$, $1 \leq i \leq n$, defined as in § 2.2, for $j = 0, 1, \ldots, p$. Note that the R_{ji}, $1 \leq i \leq n$, $1 \leq j \leq p$ are known at the beginning of the study, while the failures (the primary variate) occur sequentially over time. We consider column permutations of $\mathbf{R}_n$ such that the top row is in natural order—this reduced rank matrix is denoted by $\mathbf{R}_n^*$ and its elements by R_{ji}^*, $1 \leq i \leq n$, $0 \leq j \leq p$, where $R_{0i}^* = i$, $i = 1, \ldots, n$. This reduction process automatically generates the primary variate antiranks, which are denoted by S_{0i}, $i = 1, \ldots, n$. Consider then the $q \times (p+1)$ matrix of LRS (for the primary as well as the covariates)

$$\mathbf{L}_n = \sum_{i=1}^{n} (\mathbf{c}_i - \bar{\mathbf{c}}_n)[a_{n0}(R_{0i}), \ldots, a_{np}(R_{pi})] \tag{2.5.3}$$

where the $\mathbf{c}_i$ are defined as in § 2.2. Note that $\mathbf{L}_n$ may be rewritten as

$$\mathbf{L}_n = \sum_{i=1}^{n} (\mathbf{c}_{S_{0i}} - \bar{\mathbf{c}}_n)[a_{n0}(i), a_{n1}(R_{1i}^*), \ldots, a_{np}(R_{pi}^*)]. \tag{2.5.4}$$

Also, note that under H_0 in (2.5.2) [and given (2.5.1)], the Chatterjee–Sen rank permutation principle applies and we denote by $\mathscr{P}_n$, the conditional probability law generated by the $n!$ equally likely permutations of the columns of $\mathbf{R}_n^*$. Thus, if we let $\mathbf{S}_0^{(r)} = (S_{01}, \ldots, S_{0r})'$, for $r = 1, \ldots, n$, then as a natural extension of (2.2.7), we may consider the multivariate censored rank statistics

$$\mathbf{L}_{nr} = E_{\mathscr{P}_n}\{\mathbf{L}_n \mid \mathbf{S}_0^{(r)}\}$$

$$= \sum_{i=1}^{r} (\mathbf{c}_{S_{0i}} - \bar{\mathbf{c}}_n)[a_{nj}(R_{ji}^*) - a_{nj}^*(r), 0 \leq j \leq p] \tag{2.5.5}$$

where

$$a_{nj}^*(r) = (n-r)^{-1} \sum_{i=r+1}^{n} a_{nj}(R_{ji}^*), \quad r \leq n-1, \quad 0 \leq j \leq p, \tag{2.5.6}$$

and, conventionally, we let $a_{nj}^*(n) = 0$, $0 \leq j \leq p$. Further, we define a $(p+1) \times (p+1)$ rank covariance matrix $\mathbf{V}_n^{(r)} = ((\mathbf{v}_{njj'}^{(r)}))$; by letting for every $j, j' = 0, 1, \ldots, p$,

$$\mathbf{v}_{njj'}^{(r)} = (n-1)^{-1}\left\{\sum_{i=1}^{r} a_{nj}(R_{ji}^*)a_{nj'}(R_{j'i}^*) + (n-r)a_{nj}^*(r)a_{nj'}^*(r) - n\bar{a}_{nj}\bar{a}_{nj'}\right\}, \quad (2.5.7)$$

where the $\bar{a}_{nj}$ are defined as in (2.2.12). Then, under the permutational law $\mathscr{P}_n$, $\mathbf{L}_{nr}$ has null mean vector and covariance matrix $\mathbf{C}_n \otimes \mathbf{V}_n^{(r)}$, where $\mathbf{C}_n$ is defined by (2.2.11). Moreover, under $\mathscr{P}_n$, $\{\mathbf{L}_{nr}; 0 \leq r \leq n\}$ has also the martingale structure (by the definition in (2.5.5)). We partition

$$\underset{q \times (p+1) \quad q \times 1 \ q \times p}{\mathbf{L}_{nr} = (\mathbf{L}_{nr}^{(0)}, \mathbf{L}_{nr}^{(*)})} \quad \text{and} \quad \mathbf{V}_n^{(r)} = \begin{pmatrix} \mathbf{v}_{n00}^{(r)} & \mathbf{v}_{n0}^{(r)} \\ \mathbf{v}_{n0}^{(r)'} & \mathbf{V}_{nr}^* \end{pmatrix}, \quad (2.5.8)$$

and to eliminate the effect of the concomitant variates, we fit a regression of $\mathbf{L}_{nr}^{(0)}$ on $\mathbf{L}_{nr}^*$, and the residual vector is then

$$\mathbf{L}_{nr}^0 = \mathbf{L}_{nr}^{(0)} - \mathbf{L}_{nr}^{(*)}(\mathbf{V}_{nr}^*)^- \mathbf{v}_{n0}^{(r)'}. \quad (2.5.9)$$

Then, we have from the above

$$E_{\mathscr{P}_n}\mathbf{L}_{nr}^0 = \mathbf{0} \quad \text{and} \quad E_{\mathscr{P}_n}(\mathbf{L}_{nr}^0)(\mathbf{L}_{nr}^0)' = \mathbf{C}_n \cdot \mathbf{v}_{n*}^{(r)}, \quad (2.5.10)$$

where

$$\mathbf{v}_{n*}^{(r)} = \mathbf{v}_{n00}^{(r)} - \mathbf{v}_{n0}^{(r)}(\mathbf{V}_{nr}^*)^- \mathbf{v}_{n0}^{(r)'}. \quad (2.5.11)$$

With this, as in (2.2.14), we may consider the statistic

$$\mathscr{L}_{nr}^0 = \{(\mathbf{L}_{nr}^0)'\mathbf{C}_n^{-1}(\mathbf{L}_{nr}^0)\}/\mathbf{v}_{n*}^{(r)}. \quad (2.5.12)$$

In a PCS, at the successive failures, one computes the $\mathscr{L}_{nk}^0$ ($k \leq r$), where r is some pre-fixed positive integer ($\leq n$), and hence, as in § 2.4, we want to incorporate the partial sequence $\{\mathscr{L}_{nk}^0; k \leq r\}$ in the formulation of a time-sequential testing procedure. The procedure is very similar to the one described after (2.4.21), where we need to replace the $\mathscr{L}_{nk}$ by $\mathscr{L}_{nk}^0$, and for large n, to start testing at the k_nth failure, where k_n satisfies a condition similar to (2.4.20). To make this point clear, we may remark that we might have defined k_n by $\mathbf{v}_{n*}^{(k_n)}/\mathbf{v}_{n*}^{(r)} \geq \varepsilon > 0$. But $\mathbf{v}_{n*}^{(r)}$ (or even $\mathbf{v}_{n*}^{(n)}$) are not known at the beginning of the study (or at the k_nth failure), and hence, from the operational point of view, this definition is not of much help. On the other hand, by (2.5.11), $\mathbf{v}_{n*}^{(r)} \leq \mathbf{v}_{n00}^{(r)} = A_{n,r}^{02}$, where A_{nr}^{02} is defined as in (2.2.13), but, for the scores $a_{n0}(i)$, $1 \leq i \leq n$. Hence, $\mathbf{v}_{n*}^{(r)}/\mathbf{v}_{n*}^{(r)} \geq \mathbf{v}_{n*}^{(k)}/A_{n,r}^{02} \forall k$. Hence, if we define k_n by $\mathbf{v}_{n*}^{(k_n)}/A_{n,r}^{02} \geq \varepsilon > 0$, then we have a lower bound for ε, and hence, we end up with a somewhat conservative test.

The invariance principle in Theorem 2.3.1 has been extended to this multivariate situation, and that provides an invariance principle for the $\mathscr{L}_{nk}^0$, similar to (2.3.18)–(2.3.19). These details are provided in Sen (1979b), (1981c). The properties of this time-sequential ANOCOVA test are then very similar to those of the test in (b) of § 2.4. From the mathematical manipulations point of view, this, of course, is much more involved than the ones in § 2.4. We may conclude this section with the remark that in this development we do not need to make any linearity assumption on the effects of the concomitant variates on the primary variate (which is usually needed in the classical parametric model), nor do we need to consider the *proportional hazard model* wherein the covariates are incorporated through the hazard functions in a very special way which may not really hold in the usual case. We shall elaborate this point more in the last section.

2.6. Staggered entry and random withdrawals. It is quite common in a follow-up study that all the subjects may not simultaneously enter into the scheme at a common entry-point; they may enter into the schemes in "batches" or in some other (possibly, random) pattern. Further, withdrawals (*dropouts*) from the scheme may occur due to various causes. Let us consider the situation in which the subjects enter the scheme at (possibly) different points of time and are subjected to dropouts too. Let e_i be the time point at which the ith subject enters the study and suppose that it continues to be in the scheme until the failure occurs (at some time point $T_i = E_i + x_i^0$) or it drops out of the scheme (at time $e_i^* = e_i + Y_i$), which ever occurs first, for $i \geq 1$. Thus, X_i^0 and Y_i are the actual failure and withdrawal times, respectively, and the observable r.v. are

$$X_i = \min \{X_i^0, Y_i\}, \quad i \geq 1. \tag{2.6.1}$$

Note that because of the staggered entry plan, the X_i are not necessarily observed in order, even if there is no withdrawal. We denote by n_t, the *cumulative sample size*, the number of entries into the scheme prior to the time point t. If all the entry points are distinct, then n_t assumes all integer values between 1 and n, the target sample sizes, as t moves away from the initial entry point to ∞ and n_t is nondecreasing in t. On the other hand, if the units are admitted in l (≥ 1) batches at the time points $t_1 < \cdots < t_l$, there being n_j^* units in the jth batch ($1 \leq j \leq l$), then $n = n_1^* + \cdots + n_l^*$ and

$$n_t = n_{t_j} = n_1^* + \cdots + n_j^* \quad \text{for } t \in [t_j, t_{j+1}), \quad 0 \leq j \leq l, \tag{2.6.2}$$

where $t_0 = 0$ and $t_{l+1} = +\infty$. For any censoring point t^* ($>t_1$), we need to take into account the set of entry points $\{e_i : e_i \leq t^*\}$ (or in the batch-arrival model $\{t_1, \ldots, t_s\}$ where $t_s \leq t^* < t_{s+1}$). Then, for the censoring point t^*, the n_j^* unit entering at time point t_j have an *exposure period* $t^* - t_j$, for $j = 1, \ldots, s$. This differential pattern of the exposure periods of the subjects for the different batches introduces additional complications in carrying out a valid time-sequential testing procedure. To illustrate the procedure, we consider first the case where there is no withdrawal, so that $X_i = X_i^0$, for every $i \geq 1$. Later on, we will incorporate the dropouts in a suitable manner.

Let $n(t^*, u)$ ($=\sum_{i=1}^n I(e_i \leq t^* - u)$) be the total number of units entering the scheme on or before the time point $t^* - u$, where $t^* \geq t_1 = \min \{e_i : i = 1, \ldots, n\}$

and $0 \leq u \leq t^* - t_1$. Let then $r(t^*, u)$ be the number of failures among these n (t^*, u) units with actual failure times less than or equal to u (the remaining $n(t^*, u) - r(t^*, u)$ have failure times $> u$). Then, with respect to the sample size $n(t^*, u)$ and censoring number $r(t^*, u)$, we can, as in (2.2.8)–(2.2.9), compute

$$\mathbf{L}_{n(t^*,u)r(t^*,u)} \quad \text{for every } 0 \leq u \leq t^*, \quad t^* \geq t_1. \tag{2.6.3}$$

The definitions in (2.2.11)–(2.2.13) may also be directly adapted to this scheme, and parallel to (2.2.14), we have then

$$\mathcal{L}_{n(t^*,u)r(t^*,u)}, \quad 0 \leq u \leq t^*, \quad t^* \geq t_1. \tag{2.6.4}$$

Thus, as a natural extension of the scheme in (2.2.16), here we need to take into account the set of statistics in (2.6.4) in formulating a stopping rule on which the time-sequential procedure will rest. Basically, if any time point $t(>t_1)$, for the first time, for some u $(0 < u < t)$, $\mathcal{L}_{n(t,u)r(t,u)}$ is significantly large, we stop the study at that point of time along with the rejection of H_0. If no such t $(\leq T$, target endpoint of the study) exists, then the study is curtailed at the preplanned point of time T and the null hypothesis is accepted.

For studying the properties of this staggered entry time-sequential procedure, we need to consider some generalizations of the results in § 2.3 (due to Sen (1976a)). These we consider first. For every $m \geq 1$ and $k: 0 \leq k \leq m$, we define $\mathbf{L}_{mk}$ as in (2.2.8). Also, we define $\mathbf{C}_m$, A_{mk}^2 etc., all as in § 2.2. Note that $\mathbf{C}_n - \mathbf{C}_m$ is p.s.d. for every $n \geq m$. Then, for every n (≥ 1), we define $n(s)$, $s \in [0, 1]$, by letting

$$n(s) = \max\{m : \text{Trace } \mathbf{C}_m \leq s \text{ Trace } \mathbf{C}_n\}, \quad 0 \leq s \leq 1. \tag{2.6.5}$$

Also, as in (2.3.7), we define

$$k_n(s, t) = \max\{k : A_{n(s),k}^2 \leq t A_{n(s)}^2\}, \quad 0 \leq t \leq 1, \quad 0 \leq s \leq 1. \tag{2.6.6}$$

On the unit square $I^2 = [0, 1] \times [0, 1]$, consider a stochastic process $\mathbf{W}_n^* = \{\mathbf{W}_n^*(s, t); (s, t) \in I^2\}$, by letting

$$\mathbf{W}_n^*(s, t) = A_n^{-1} \mathbf{C}_n^{-1/2} \mathbf{L}_{n(s)k_n(s,t)}, \quad (s, t) \in I^2. \tag{2.6.7}$$

Also, let $\mathbf{W}^* = \{\mathbf{W}^*(s, t); (s, t) \in I^2\}$ be $q(\geq 1)$ independent copies of a Brownian sheet on $[0, 1]^2$. Then, for $q = 1$, the following functional central limit theorem is due to Sen (1976a):

$$\text{Under } H_0, W_n^* \text{ converges weakly to } W^*. \tag{2.6.8}$$

For $q > 1$, the same result holds under an additional *growth condition*:

$$\lim_{m \to \infty} m^{-1} \mathbf{C}_m = \mathbf{C}_0 \text{ is p.d.}; \tag{2.6.9}$$

we may refer to Sinha and Sen (1983) in this context. (2.6.9) holds typically in the multi-sample case described after (2.2.1). As such, if we consider q-dimensional Bessel sheets $B_q^* = \{B_q^*(s, t) : (s, t) \in I^2\}$ by letting

$$[B_q^{**}(s, t)]^2 = [\mathbf{W}^*(s, t)]'[\mathbf{W}^*(s, t)], \quad (s, t) \in I^2, \tag{2.6.10}$$

and define $B_{nq}^* = \{B_{nq}^*(s, t) : (s, t) \in I^2\}$ by letting

$$[B_{nq}^*(s, t)]^2 = [\mathbf{W}_n^*(s, t)]'[\mathbf{W}_n^*(s, t)]$$
$$= A_n^{-2}\{\mathbf{L}'_{n(s)k_n(s,t)}\mathbf{C}_n^{-1}\mathbf{L}_{n(s)k_n(s,t)}\}, \qquad (2.6.11)$$

for $(s, t) \in I^2$, then we have under H_0 and (2.6.9),

$$B_{nq}^* \text{ converges weakly to } B_q^{**}. \qquad (2.6.12)$$

As such, if $\delta_{q\alpha}^*$ be the upper $100\alpha\%$ point of the distribution of $\sup\{B_q^{**}(s, t) : (s, t) \in I^2\}$, then under H_0 and (2.6.9),

$$P\left\{\max_{m \leq n} \max_{k \leq m} A_n^{-2}\mathbf{L}'_{mk}\mathbf{C}_n^{-1}\mathbf{L}_{mk} \leq \delta_{q\alpha}^* \mid H_0\right\} \approx 1 - \alpha. \qquad (2.6.13)$$

This asymptotic result provides the basic tool for the formulation of time-sequential nonparametric tests for the staggered entry plan. With a view to incorporating censored plans, we proceed as in Majumdar and Sen (1978b) and consider the following. Let I^* be a subset of I^2. Then, by virtue of (2.6.12), under H_0,

$$\sup\{B_{nq}^*(s, t) : (s, t) \in I^*\} \underset{\mathscr{D}}{\to} \sup\{B_q^{**}(s, t) : (s, t) \in I^*\}. \qquad (2.6.14)$$

Typically, in a censored (or truncated) experiment with a staggered entry plan, I^* relates to the lower left-hand part of I^2; see Fig. 2.6.1, for example.

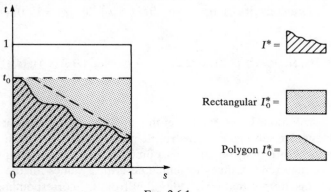

FIG. 2.6.1.

The upper boundary of I^* depends on the entry as well as failure patterns. While, in some cases, the entry pattern may be set in advance, the failure pattern is generally stochastic, and hence, in general I^* is a stochastic subset of I^2. The cumulative sample sizes at different time points and the number of failures corresponding to the different exposure periods (during the tenure of the study) determine the shape of I^* (by (2.6.5) and (2.6.6)). However, in many cases, it may be possible to provide close upper bounds to these cohort failures (from previous or independent studies) and thereby to close I^* by a nonstochastic I_0^*, with probability very close to 1. In any case, if we choose

$I_0^* = \{(s, t) : 0 \leq s \leq 1, 0 \leq t \leq t_0\}$, where t_0 corresponds to the initial batch failure rate, then we have a rectangular I_0^* containing I^*. Thus, if we use (2.6.14) for I_0^* and compute the critical values for the Bessel (Brownian) sheet over I_0^*, we may employ these for the statistical monitoring of the trial in a PCS with staggered entry. In view of the fact that $I^* \subset I_0^*$, these critical values may be somewhat larger than the actual ones, and this will result in a somewhat conservative property, though this effect will be very minor if I_0^* is very close to I^*. In the case of a "batch arrival model", we may take I_0^* as a polygon with straight line boundaries on the left and bottom and with descending steps on the other sides.

The problem of determining the percentile points of the distribution of the supremum of B_q^{**} over I_0^* (or I^2, even) by analytical methods has been a challenging issue for quite some time. Though very (close) bounds for these are available in the literature, closed expressions for the exact distributions are not precisely known. Fortunately, these critical values can be obtained quite satisfactorily by simulating Brownian (or Bessel) sheets, identifying the appropriate I_0^* and deriving the percentile points by the usual Monte Carlo methods. This approach has been pursued in Majumdar and Sen (1978b) and Sinha and Sen (1982), (1983) and their simulation studies are in fairly good agreement with the available bounds. These simulation studies also rest on the weak invariance principle for independent random variables with vector indices $(i, j) : 1 \leq i \leq N, 1 \leq j \leq N$, for which theoretically the Brownian (or Bessel) sheet approximations are valid. In passing, we may remark that for the case of rectangular subregions $\{(s, t) : 0 \leq s \leq p_1, 0 \leq t \leq p_2\}$, $0 < p_1 \leq 1$, $0 \leq p_2 \leq 1$, we have

$$\sup \{B_q^{**}(s, t) : 0 \leq s \leq p_1, 0 \leq t \leq p_2\} \underset{\mathcal{D}}{=} \sqrt{p_1 p_2} \cdot \sup \{B_q^{**}(s, t); (s, t) \in I^2\}$$

$$(2.6.15)$$

so that the distribution theory for the entire square (I^2) provides the same for all such rectangular subregions. This result may be quite useful in the batch arrival model where I_0^* may be taken as a union of some rectangular subsets of I^2. Finally, we may remark that the simulated critical values for the Brownian (or Bessel) sheets are not that widely different from the corresponding ones for the Brownian (or Bessel) processes, and hence, adoption of this statistical monitoring in a PCS with staggered entry does not really lead to too much conservatism in terms of loss of power. On the other hand, it provides a valid statistical analysis with the provision of an early stopping, subject to a given risk of making an incorrect decision.

Dropouts or withdrawals are quite common in followup studies. It is possible to accommodate such dropouts in a PCS under suitable assumptions. With the notation in (2.6.1), if we assume that the failure time X_i^0 and the withdrawal time Y_i are mutually stochastically independent and further the Y_i are i.i.d.r.v.

with a d.f. G, then

$$H_i(x) = P\{x_i \leq x\} = 1 - P\{X_i > x\}$$
$$= 1 - P\{X_i^0 \geq x, Y_i \geq x\}$$
$$= 1 - \{1 - F_i(x)\}\{1 - G(x)\} \quad \forall i \geq 1, \qquad (2.6.16)$$

so that under $H_0 : F_1 = \cdots = F_n = F$, $H_1 = \cdots = H_n = H$ $(= 1 - (1 - F)(1 - G))$, and hence, the testing schemes considered in §§ 2.3, 2.4, 2.5 and earlier in this section are all applicable to this random withdrawal model. Thus, from the operational point of view, no change is needed to accommodate random withdrawals in the PCS. However, we may note that by (2.6.16),

$$H_i(x) - H_j(x) = \{1 - G(x)\}\{F_i(x) - F_j(x)\}, \qquad (2.6.17)$$

for all $i \neq j = 1, \ldots, n$, and hence, when the null hypothesis is not true, because of the fact that $1 - G(x) \leq 1$, $\forall x$, we have a damping factor in (2.6.17), which diminishes the distances between the distributions, and this may lead to some loss of power. The assumption of independence of Y_i and X_i^0 may not also be very realistic in many situations. In the negation of this assumption, the distribution-free character of the censored rank statistics may not hold. The behavior of specific test statistics may depend on the joint distribution of (X_i^0, Y_i) and when the Y_i are not i.i.d., even if they are independent, standard results such as the ones discussed in this chapter may not be applicable for them. Of course, some parametric analysis may be made using specific distributional assumptions on the (X_i^0, Y_i), but, we shall not go into the details of such parametric models.

2.7. Some miscellaneous time-sequential procedures. The procedures considered so far are primarily rank based ones. Some other time-sequential procedures, based on empirical processes, score statistics and the Cox (1972) proportional hazard model, will be briefly discussed here. Again, we concentrate mainly on the time-sequential aspect of these procedures, and, by way of references, leave the other aspects untouched.

Consider first some nonparametric procedures based on (generalized versions of) Kolmogorov–Smirnov type statistics in a PCS. To motivate these procedures, consider the two-sample problem treated after (2.2.1). Let F_{n_1} and G_{n_2} be the two-sample (empirical) d.f. In the context of life testing under PCS, one may, as in § 2.2, study the process $\{n_1 n_2 (n_1 + n_2)^{-1} (F_{n_1}(x) - G_{n_2}(x)), x \geq 0\}$, so that if at any time point t ($\leq T$, the projected endpoint of the study), this process crosses the set boundary, the experimentation is curtailed along with the rejection of the null hypothesis H_0 (of equality of the two parent d.f.'s). Otherwise, experimentation stops at the present time point T, with the acceptance of H_0. Thus, one basically needs to study the distribution of the suprema of this process over the range $[0, T]$—percentile points for these are tabulated by Koziol and Byar (1975) and Schey (1977). In view of the weak convergence of this process (under H_0) to a Brownian bridge, the asymptotic results follow directly from the distribution theory of the Brownian bridge process over $[0, F(T)]$, where $0 < F(T) \leq 1$, and this, in turn, follows from that of the Brownian motion process over $[0, F(T)/\{1 - F(T)\}]$, for which the results of Anderson (1960) are accessible. In view of the regression model in (2.2.2), we consider now the generalization of such procedures for the regression model based on weighted empirical processes. These developments are mostly due to Sinha and Sen (1979a, b), (1982), (1983).

With the same notation as in § 2.2, consider the empirical processes:

$$\mathbf{H}_n(x) = \mathbf{C}_n^{-1/2} \sum_{i=1}^{n} (\mathbf{c}_i - \bar{\mathbf{c}}_n) u(x - x_i), \qquad x \geq 0, \tag{2.7.1}$$

$$S_n(x) = n^{-1} \sum_{i=1}^{n} u(x - x_i), \qquad x \geq 0, \tag{2.7.2}$$

where $u(t)$ is 1 or 0 according as t is $\geq$ or < 0, $\mathbf{C}_n$ is defined by (2.2.11), and $\mathbf{C}_n^{-1/2}$ as in after (2.3.4). Then, under $H_0 : F_1 = \cdots = F_n = F$, both $\{(1 - F(x))^{-1} S_n(x); x \in R\}$ and $\{(1 - F(x)^{-1}\mathbf{H}_n(x); x \in R\}$ are martingales (see Sen (1981, pp. 357–358)). Let $W = \{w(t), 0 < t < 1\}$ be some suitable nonnegative and continuous weight function. Then consider a generalized Kolmogorov–Smirnov type statistic

$$K_n^* = \sup \{[\mathbf{H}_n(x)]'[\mathbf{H}_n(x)]/w^2(S_n(x)) : x \in E^* \subset R\} \tag{2.7.3}$$

where E^*, a subinterval of R, may be chosen depending on the scheme at hand. For example, if we consider the unweighted version (i.e., $w(t) \equiv 1$), then as before we may take $E^* = [0, T]$. In the other notable case, $w^2(t) = t(1 - t)$, as in § 2.3, we may restrict ourselves to $E^* = [T_1, T_2]$, where $S_n(T_1) \geq \varepsilon > 0$ and $S_n(T_2) \leq 1 - \varepsilon < 1$. In view of the martingale property mentioned before, weak invariance principles for the H_n and S_n hold: Parallel to Theorem 2.3.1, here we have the tied-down Bessel process $B_q^0 = \{B_q^0(t), 0 \leq t \leq 1\}$, where $B_q^0(t) = \{(\sum_{j=1}^{q} (\xi_j(t) - t\xi_j(1))^2)^{1/2}; 0 \leq t \leq 1\}$, where the ξ_j are defined as in (2.3.13). For $q = 1$, this reduces to a version of the Brownian bridge. In the standardized case, $B_q^{0*} = \{B_q^{0*}(t) = B_q^0(t)/\sqrt{t(1-t)} : 0 < t < 1\}$, and critical values for the level crossing of these Bessel processes have been studied by Kiefer (1959) and DeLong (1981), among others. Sinha and Sen (1982), (1983) have also considered the staggered entry plan and obtained, by simulation, the percentile points of the suprema of the corresponding Bessel sheets. In view of these developments, the time-sequential procedures may be based on the process $\{[\mathbf{H}_n^T(x)]\mathbf{H}_n(x)]/w^2(S_n(x)); x \geq 0\}$, and the theory runs parallel to the two-sample case. For local alternatives, asymptotic power properties were also studied by Sinha and Sen (1979a, b), (1982), (1983).

The procedures based on the ranks or the empirical processes do not utilize any information contained in the associated vector of ordered failures. This information may be incorporated in some *scores statistics* which we formulate as follows. These results are mainly adopted from Sen (1979a), (1981b). For the regression model in (2.2.2), if we consider the log-likelihood function for the set of r.v.'s in (2.2.5) and evaluate its derivative (vector) with respect to $\boldsymbol{\beta}$ at $\boldsymbol{\beta} = \mathbf{0}$, the corresponding score statistic (vector) is

$$\mathbf{T}_{nk} = \sum_{i=1}^{k} (\mathbf{c}_{S_{ni}} - \bar{\mathbf{c}}_n)[g(Z_{n,i}) - \bar{G}(Z_{n,k})] \qquad (2.7.4)$$

where $g(x) = -f'(x)/f(x)$ and by virtue of

$$\int g \, dF = 0, \quad \bar{G}(x) = -\{1 - F(x)\}^{-1} \int_{-\infty}^{x} g(y) \, dF(y).$$

Note that we have reparameterized (2.2.2) as $F(x - \beta_0^{(n)} - \boldsymbol{\beta}'(\mathbf{c}_i - \bar{\mathbf{c}}_n))$, where $\beta_0^{(n)} = \beta_0 + \boldsymbol{\beta}'\bar{\mathbf{c}}_n$, which accounts for the $\bar{\mathbf{c}}_n$ in (2.7.4). A natural estimator of $\bar{G}(x)$ is

$$\bar{G}_n(x) = -\{1 - F_n^*(x)\}^{-1} \int_{-\infty}^{x} g(y) \, dF_n^*(y)$$

where F_n^* is the sample d.f. This leads to the following *adapted score statistics:*

$$\mathbf{T}_{nk}^0 = \sum_{i=1}^{k} (\mathbf{c}_{S_{ni}} - \bar{\mathbf{c}}_n)[g(Z_{n,i}) - G_n(Z_{n,k})]$$

$$= \sum_{i=1}^{k} g(Z_{n,i}) \left\{ (\mathbf{c}_{S_{ni}} - \bar{\mathbf{c}}_n) + (n-k)^{-1} \sum_{j=1}^{k} (\mathbf{c}_{S_{nj}} - \bar{\mathbf{c}}_n) \right\}, \qquad (2.7.5)$$

for $k = 1, \ldots, n$. Since under H_0, $\mathbf{S}_n$ and $\mathbf{Z}_n$ are stochastically independent, by arguments similar to those following (2.2.7), for every n (≥ 1), $\{\mathbf{T}_{nk}^0; k \leq n\}$ forms a martingale sequence when H_0 holds. We may choose in this context a general class of $\{g\}$—yielding the so-called M-statistics. Choice of some robust $g(\cdot)$ would result in a robust M-test for the proposed time-sequential scheme. Whereas the $\mathbf{L}_{nk}$ in (2.2.8) are genuinely distribution-free under H_0, the $\mathbf{T}_{nk}^0$ are only conditionally (given $\mathbf{Z}_n$) distribution-free. Nevertheless, by virtue of the martingale property, invariance principles (similar to those in § 2.3) hold for these adapted scores statistics as well. As such, both the time-sequential procedures (a) and (b) in § 2.4 can be based on these statistics. Therefore, the details are omitted.

The rank procedures in § 2.2 through 2.5 are based on the minimal assumptions on the underlying distributions. Cox (1972) has introduced a *quasi-nonparametric model* where the *hazard functions* for a given set of covariates are proportional to a basic (unknown) hazard function and these proportionality factors are sole functions of these covariates (of specified forms involving unknown (regression) parameters). For this reason, these are termed *proportional hazard models* (PHM). For such a PHM, Cox (1972), (1975) has justified (rather heuristically) the adaptability of the *partial likelihood approach* and proposed the usual *scores statistics* relating to these partial likelihood functions for the formulation of suitable test statistics. For large sample sizes, under quite

general regularity conditions, the usual chi-square distributional approximations for these scores statistics work out well (when the basic assumption of
PHM holds). The very adaption of the partial likelihood function leads to the
basic result that under (PHM and the) null hypothesis, the related scores
statistics sequence form a martingale—this has been exploited in Sen (1981a)
in the formulation of some time-sequential testing procedures. Invariance
principles for these partial likelihood scores (very similar to those for the LRS
in § 2.3) hold under quite general regularity conditions, and hence, time-
sequential procedures (parallel to (a) and (b) in § 2.4) work out well for such
quasi-nonparametric model. However, one should keep in mind that in this
setup, the PHM assumption plays a vital role and negation of this basic
assumption (of PHM) may lead to considerable loss of efficiency of the
time-sequential procedures for which the test statistics are structured on the
PHM. Thus, compared to the rank procedures, these may not be very robust
against possible departures from the PHM. Further, in practice, if the
covariates are spanned on a wider domain, the PHM may not be very
appropriate—a change in the unit of measurements for these covariates may
lead to a different model and possibly different statistical conclusions (for
example, a covariate measured on the ordinary or logarithmic scale may
change the PHM and may lead to different conclusions). The rank ANOCOVA
procedures described in Section 2 do not depend on the units of measurement
of the covariates, and hence, are robust in this sense. On the other hand, if the
PHM holds, the allied time-sequential procedures may be more efficient than
the rank procedures in many cases (viz. Sen (1984a)). Thus, the PHM may be
adapted in some cases with advantages.

This partial likelihood approach based on the PHM has paved the way for
some fruitful methodological research in survival analysis, where the Cox
model has been extended for certain class of *counting processes*. Consider a
stochastic process $X = \{X_t, t \in R^+\}$, $R^+ = [0, \infty)$, defined on a probability space
$(\Omega, \mathscr{F}, P)$. Then $\mathscr{F}_t^X = \sigma(X_s, s \in [0, t])$, a sub-sigma-field of $\mathscr{F}$, $t \geq 0$, is called the
internal history of the process X_t, $t \geq 0$. A family $\{\mathscr{F}_t, t \in R^+\}$ of sub-sigma fields
of $\mathscr{F}$ is called a *history* of X if it is increasing (in $t \in R^+$) and

$$\mathscr{F} \supseteq \mathscr{F}_t^X \quad \forall t \in R^+; \tag{2.7.6}$$

$\{X_t, t \in R^+\}$ is then said to be *adapted to* $\{\mathscr{F}_t, t \in R^+\}$. Consider events occurring
at time points $\tau_1 \leq \tau_2 \leq \cdots \leq \tau_n \leq \cdots$ on R^+, and let $N(t)$ be the number of
events occurring in $[0, t]$, $t \in R^+$. Then $\{N(t), t \in R^+\}$ is a counting process,
where $N(t)$ is nonnegative, integer valued and $N(t)$ is $\nearrow$ in $t \in R^+$. Further,
assume that τ_j has the distribution function F_j which admits of a continuous
p.d.f. f_j, every $j \geq 1$. Then, note that

$$H(t) = EN(t) = \sum_{j=1}^{\infty} F_j(t), \qquad t \in R^+, \tag{2.7.7}$$

so that the *point process intensity function* $h = \{h(t), t \in R^+\}$ is defined by

$$h(t) = H'(t) = \sum_{j=1}^{\infty} f_j(t), \qquad t \in R^+. \tag{2.7.8}$$

Note that $h(t)$ is locally integrable and is the density for the intensity measure. At this stage, we may quote the Watanabe (1964) theorem:

Let $\{N_t, t \in R^+\}$ be a point process adapted to the history $\{\mathscr{F}_t, t \in R^+\}$, and let $\lambda = \{\lambda(t), t \in R^+\}$ be a locally integrable, nonnegative, measurable function, such that

$$N_t - \int_0^t \lambda(s) \, ds \text{ is a } \mathscr{F}_t\text{-martingale,} \qquad t \in R^+. \tag{2.7.9}$$

Then N_t is an $\mathscr{F}_t$-process with the intensity $\lambda(t)$, $t \in R^+$, that is, for all $0 \le s \le t < \infty$, $N_t - N_s$ is a Poisson r.v. with the parameter $\int_s^t \lambda(u) \, du$, independent of $\mathscr{F}_s$.

A multichannel extension of this result is as follows:

Let $\mathbf{N}_t = (N_t(1), \ldots, N_t(k))'$ be a $k \, (\ge 1)$-variate point process adapted to some history $\mathscr{F}_t, t \in R^+$, and suppose that

$$\mathbf{N}_t - \int_0^t \boldsymbol{\lambda}(u) \, du \text{ is an } \mathscr{F}_t\text{-martingale,} \qquad t \in R^+, \tag{2.7.10}$$

where $\boldsymbol{\lambda}(t) = (\lambda_1(t), \ldots, \lambda_k(t))', t \in R^+$, is a nonnegative, locally integrable measurable function. Then the $N_t(i) \ (1 \le i \le k)$ are independent $\mathscr{F}_t$-Poisson processes with intensities $\lambda_i(t), 1 \le i \le k$, respectively $(t \in R^+)$. (Note that the $\lambda_i(t)$ must be deterministic.)

Nonparametric inference for some counting processes was first considered by Aalen (1978), (1980); further useful contributions to this interesting area are due to Anderson and Gill (1982), Anderson, Borgan, Gill and Keiding (1982), Johansen (1983), and others. Consider a $k \, (\ge 2)$-dimensional counting process $\mathbf{N} = \{(N_1(t), \ldots, N_k(t))', t \in [0, 1]\}$ with the intensity process $\boldsymbol{\lambda} = (\lambda_1(t), \ldots, \lambda_k(t))', t \in [0, 1]\}$ and assume that there exist an $(\mathscr{F}_t\text{-})$ adapted stochastic process $\mathbf{Y} = \{(Y_1(t), \ldots, Y_k(t))', t \in [0, 1]\}$ and functions $\boldsymbol{\alpha} = \{(\alpha_1(t), \ldots, \alpha_k(t))', t \in [0, 1]\}$, such that

$$\lambda_i(t) = \alpha_i(t) Y_i(t), \qquad t \in [0, 1], \quad i = 1, \ldots, k. \tag{2.7.11}$$

In this setup, the process $\mathbf{Y}$ does not depend on a parametric model (compare to the unknown hazard function in the PHM) while the α_i are deterministic and are parametric in nature. Typically, one wants to test the hypothesis

$$H_0 : \alpha_1(t) = \cdots = \alpha_k(t), \qquad t \in [0, 1]. \tag{2.7.12}$$

(For $k = 1$, a similar model works out where one wants to test for $H_0 : \alpha(t) = \alpha_0(t), t \in [0, 1]$, for some specified α_0.) Then (2.7.10) extends to such proportional models (PM) (in (2.7.11)), and based on this (local) martingale structure, invariance principles work out in a natural way. Thus, time-sequential procedures can be based on the counting process $\{\mathbf{N}(t), t \in [0, 1]\} = \mathbf{N}$ under the PM in (2.7.11) and asymptotic results parallel to those in earlier sections follow. Like the Cox (1972) model, for such counting processes too, the PM in (2.7.11) plays a vital role and, in the negation of this assumption, the procedures may be quite inefficient.

Throughout this chapter, we have considered PC tests for suitable null hypotheses (of invariance) against some global alternatives. In a variety of situations in testing for a null hypothesis of homogeneity (or invariance), one may confront some *restricted alternatives*. For example, in the several sample problem, the null hypothesis of interest is the homogeneity of all the distributions, and one may be interested in the set of *ordered alternatives* (where the distributions are ordered). Similarly, in the Cox proportional hazard model, the null hypothesis of interest may be $\boldsymbol{\beta} = \mathbf{0}$, against the *orthant alternative*: $\boldsymbol{\beta} > \mathbf{0}$. In such a case of restricted alternatives, the tests considered earlier are valid, but may not be very efficient. More efficient procedures (taking into account the restrictions on the parameters under testing) may be obtained by using the Roy (1953) *union-intersection principle* along with the current PCS methodology. In this context too, the invariance principles considered earlier play the basic role. Unlike the case of approximations by Brownian motion, Brownian bridge or Bessel processes, treated earlier, we confront the case of some *Bessel bar processes* (which are the natural stochastic processes related to the usual *chi-bar* distributions). For the Cox model as well as linear rank statistics, invariance principles for such restricted alternative PC test statistics have been studied in Sen (1984g), (1984h), and these are fruitfully incorporated in the study of the asymptotic properties of suitable PC tests for such restricted alternatives. There is a genuine need to develop analytical methods for numerical evaluation of distributions arising in this context. The results of DeLong (1981) need to be extended for such Bessel bar processes too. The prospect for simulation studies looks very bright, and, on a limited basis, some of these studies have already been carried out successfully. Estimation of parameters in the conventional location or regression models or the regression parameters in the proportional hazard model under PCS is both theoretical and applied interests. For a general class of parametric models, progressively truncated maximum likelihood estimators (PTMLE) have been considered in detail by Inagaki and Sen (1985) and the asymptotic theory of PTMLE has been systematically developed there. For the nonparametric case, however, the developments are rather spotty and piecemeal. This is largely due to a technical difficulty with the alignment principle (underlying the R- and M-estimation theory) under PCS. Thus, alternative procedures for estimation following nonparametric PC tests remain to be explored, and we pose these as open problems.

CHAPTER 3
Change-Point Nonparametrics

3.1. Introduction. Let $\{X_i; i \geq 1\}$ be a sequence of independent r.v., taken at ordered time points $\{t_i; i \geq 1\}$, with d.f. $\{F_i; i \geq 1\}$, all belonging to a common family $\mathcal{F}$. In a conventional nonsequential model, given $n(\geq 1)$, based on the assumption that $F_1 = \cdots = F_n = F$, one draws statistical inferences on F (or its parameters). There are, however, problems (typically arising in continuous sampling inspection plans, see Page (1957)) in which a change of the d.f. (or parameters) may occur at some unknown time point τ (the *change-point*), where $\tau \in (t_1, t_n)$. Hence, one may desire to test the null hypothesis of equality of $F_1, \ldots, F_n$ against the composite alternatives that $F_1 = \cdots = F_q \neq F_{q+1} = \cdots = F_n$ for some q (unknown); $1 \leq q \leq n-1$. A similar situation arises in a *sequential detection problem* (cf. Page (1954) and Shiryaev (1963), (1978)) where one assumes that for some q (possibly, $+\infty$), $X_1, \ldots, X_q$ are i.i.d.r.v. with the d.f. F_1 and X_{q+j}, $j \geq 1$ are i.i.d.r.v. with the d.f. F_{q+1} ($\neq F_1$), and the problem is to raise an alarm if $q < \infty$. Thus, one would like to choose a *stopping rule* N such that (i) if actually $q < \infty$, then $(N-q)^+$ ($= \max (0, N-q)$) should be as small as possible (*quickest detection*), while, (ii) if $q = +\infty$, the probability of a *false alarm* (i.e., $N < \infty$) should be small. Often, for (i) one considers $E[(N-q)^+ | q < \infty]$ (as small as possible) and for (ii) $E[N | q = +\infty]$ (as large as possible), though some other criteria may be used as well.

Whereas in a sequential detection problem, one has a proper sequential procedure based on a stopping rule, in a change-point problem (with n specified in advance), the tests need not be of sequential nature. The sequential detection problem may be more appropriate in statistical quality control, but, in view of the periodic inspections and adjustments usually prescribed in such setups, the change-point problem remains equally appealing in such a case. However, within each inspection period (relating to a preassigned value of n), the test for the change-point may be made recursively, so that an early stopping is admissible. Thus, a *quasi-sequential procedure* seems to be quite appropriate.

In this chapter, we shall mainly confine ourselves to such quasi-sequential procedures in the nonparametric case. In the parametric case, Brown, Durbin and Evans (1975) have dealt with tests for the *constancy of regression relations over time* based on *recursive residuals*, and these tests are of quasi-sequential nature. Invariance principles for recursive residuals were studied by Sen (1982a) with a view to incorporating them in the change-point problem without making explicit assumption on the basic form of the underlying d.f. In the nonparametric case, *recursive ranking* has been incorporated in the change-point problem (location model) by Bhattacharya and Frierson (1981). In the more general linear model case, a natural extension of the recursive ranking is the *recursive aligned-ranking*, considered by Sen (1983a). *M*-tests for change-points based on recursive residuals have also been considered by Sen (1983b). For a general class of estimable parameters, tests for change-points based on *recursive U-statistics* have also been considered by Sen (1982c). We shall mainly review the general theory of these recursive tests for change-points, with some references to other tests. Again, our main contention is to show that the usual tools of sequential nonparametrics may be adapted to handle the change-point problem in a natural manner.

Section 3.2 is devoted to the tests based on recursive *U*-statistics; the asymptotic theory of the tests is discussed here. Section 3.3 deals with the tests based on rank order statistics and recursive aligned ranking is incorporated in the scheme. Section 3.4 is devoted to the *M*-tests based on recursive residuals. The last section contains some general remarks and useful discussions.

3.2. Tests based on recursive U-statistics. Consider a Borel measurable *kernel* $g(x_1, \ldots, x_m)$, symmetric in its m (≥ 1) arguments, such that

$$\theta(F) = E_F\{g(X_1, \ldots, X_m)\}, \text{ for every } F \text{ belonging to a class } \mathcal{F}. \quad (3.2.1)$$

Then, $\theta(F)$ is an *estimable parameter of degree m*. Under the null hypothesis H_0 that $F_1, \ldots, F_n$ are all equal to some (unknown) F, for every $n \geq m$, a symmetric, unbiased and optimal estimator of $\theta(F)$ is

$$U_n = \binom{n}{m}^{-1} \sum_{1 \leq i_1 < \cdots < i_m \leq n} g(X_{i_1}, \ldots, X_{i_m}). \quad (3.2.2)$$

We may refer to Hoeffding (1948) for a very elaborate study of the distributional properties of these *U*-statistics. The sample mean is a special case of U_n when $m = 1$ and $g(x) = x$. For normal d.f., tests for change-points based on the sample means have been studied by Chernoff and Zacks (1964) and others. These are based either on the residuals $\{\bar{X}_k - \bar{X}_n; k = 1, \ldots, n-1\}$ (where $\bar{X}_k = k^{-1} \sum_{i=1}^{k} X_i, k \geq 1$) or on the recursive residuals $\{X_{k+1} - \bar{X}_k; k = 1, \ldots, n\}$. Note that for every $k: 1 \leq k \leq n-1$,

$$\bar{X}_k - \bar{X}_n = n^{-1}(n-k)\{\bar{X}_k - _{n-k}\bar{X}\}, \qquad _{n-k}\bar{X} = (n-k)^{-1} \sum_{j=k+1}^{n} X_j, \quad (3.2.3)$$

so that one can also consider a pseudo-two-sample approach where $X_1, \ldots, X_k$ are treated to be from a first population and $X_{k+1}, \ldots, X_n$ from a second one; since the change-point is unknown, one needs to consider all possible (i.e., $(n-1)$) such pairs of subsamples. This approach has been extended to the case of *U*-statistics in Sen (1982c). To extend the recursive residuals to the general case, we rewrite

$$X_{k+1} - \bar{X}_k = (k+1)\{\bar{X}_{k+1} - \bar{X}_k\}, \qquad k \geq 1. \quad (3.2.4)$$

This leads us to consider the following sequence:

$$(k+1)\{U_{k+1} - U_k\}, \qquad k = m, m+1, \ldots, n. \quad (3.2.5)$$

To simplify this further, we introduce the *recursive U-statistics*

$$U_k^* = \binom{k-1}{m-1}^{-1} \sum_{1 \le i_1 < \cdots < i_{m-1} \le k-1} g(X_{i_1}, \ldots, X_{i_{m-1}}, X_k), \qquad k \ge m.$$
(3.2.6)

Note that by definition

$$(k+1)(U_{k+1} - U_k) = m(U_{k+1}^* - U_k), \qquad k \ge m.$$
(3.2.7)

We employ the cumulative sums (CUSUM's) of the residuals in (3.2.7) for the testing problem. Towards this, we define

$$W_k = \sum_{i=m+1}^{k} (U_i^* - U_{i-1}), \quad k \ge m+1, \qquad W_k = 0 \quad \text{for } k \le m.$$
(3.2.8)

Note that (cf. Hoeffding (1984)) for every $n \ge m$,

$$E[U_n - \theta(F)]^2 = \binom{n}{m}^{-1} \sum_{c=1}^{m} \binom{m}{c}\binom{n-m}{m-c}\zeta_c = n^{-1}m^2\zeta_1 + O(n^{-1}), \quad (3.2.9)$$

where the ζ_c are nonnegative (and nondecreasing in c), and are defined by

$$\zeta_c = \text{cov}[g(X_1, \ldots, X_m), g(X_{m-c+1}, \ldots, X_{2m-c})], \qquad c = 0, 1, \ldots, m, \quad \zeta_0 = 0.$$
(3.2.10)

For every $n \, (\ge m+1)$ and $i : 1 \le i \le n$, we let

$$U_{ni} = \binom{n-1}{m-1}^{-1} \sum_{\substack{1 \le i_2 < \cdots < i_m \le n \\ (i_j \ne i)}} g(X_i, X_{i_2}, \ldots, X_{i_m}), \qquad (3.2.11)$$

and let

$$s_n^2 = (n-1)^{-1} \sum_{i=1}^{n} (U_{ni} - U_n)^2, \qquad n \ge m+1.$$
(3.2.12)

The tests for the change-point are based on the statistics

$$D_n^+ = (n-m)^{-1/2} \max\{W_k/\tilde{s}_k : m < k \le n\} \quad \text{(one-sided)}, \qquad (3.2.13)$$

$$D_n = (n-m)^{-1/2} \max\{|W_k|/\tilde{s}_k : m < k \le n\} \quad \text{(two-sided)}, \qquad (3.2.14)$$

where $\tilde{s}_k = \max\{s_k, k^{-1/2}\}$, $k \ge m+1$. If the ζ_c are known, then in (3.2.13)–(3.2.14), one may also replace the $\tilde{s}_k$ by $\zeta_1^{1/2}$. If for a given level of significance $\alpha(0 < a < 1)$, $\delta_{n,\alpha}^+$ and $\delta_{n,\alpha}$ be respectively the upper $100\alpha\%$ point of the distribution of D_n^+ and D_n under H_0, then, operationally, the control chart adapted test procedure consists in looking at the CUSUM's $W_k/\tilde{s}_k$, $k \ge m+1$ and rejecting H_0 if for some $k : m < k \le n$, $W_k/\tilde{s}_k$ (or $|W_k|/\tilde{s}_k$) exceeds $(n-m)^{1/2}\delta_{n,\alpha}^+$ (or $(n-m)^{1/2}\delta_{n,\alpha}$). Thus, we have a quasi-sequential procedure, where a stopping rule (N) is defined properly $(m < N \le n)$ in terms of the W_k and $\tilde{s}_k$, $m \le k \le n$. The basic difference between this quasi-sequential procedure and the parallel one in a sequential detection problem is that here N is bounded from above by n, while it could be $+\infty$ in the other case. From the operational point of view, one needs to determine the critical values $\delta_{n,\alpha}^+$ and $\delta_{n,\alpha}$, and, towards this, the invariance principles for recursive U-statistics and the strong consistency of the s_n play a vital role.

For every $n\,(\geq m)$, we define a stochastic process $Z_n = \{Z_n(t); t \in [0, 1]\}$ by letting $Z_n(k/n) = k[U_k - \theta(F)]/\{m(n\zeta_1)^{1/2}\}$, $k = m, \ldots, n$; $Z_n(k/n) = 0$, $k < m$, and $Z_n(t) = Z_n(k/n)$ for $k/n \leq t < (k+1)/n$, $k \geq 0$. Then Z_n belongs to the $D[0, 1]$ space equipped with the Skorokhod J_1-topology. By making use of the Hoeffding (1961) decomposition of U-statistics along with the reverse martingale property of the same, Miller and Sen (1972) have shown that whenever $0 < \zeta_1 \leq m^{-1}\zeta_m < \infty$,

$$Z_n \underset{\mathscr{D}}{\Rightarrow} Z, \quad \text{in the } J_1\text{-topology on } D[0, 1], \text{ as } n \to \infty, \tag{3.2.15}$$

where $Z = \{Z(t); t \in [0, 1]\}$ is a standard Wiener process on $[0, 1]$. Note that by (3.2.7) and (3.2.8), for every $k \geq m + 1$,

$$(n\zeta_1)^{-1/2} W_k = Z_n(k/n) - \sum_{i=m}^{k-1} i^{-1} Z_n(i/n) - m Z_n(m/n). \tag{3.2.16}$$

As such, if we define $Z_n^* = \{Z_n^*(t); t \in [0, 1]\}$ by letting $Z_n^*(t) = Z_n^*(k/n)$ for $k/n \leq t < (k+1)/n$, $k \geq 0$, where $Z_n^*(k/n) = (n\zeta_1)^{-1/2} W_k$, $k \geq m + 1$; and 0 for $k \leq m$, then, by using (3.2.16) and proceeding as in Sen (1982c), it follows that under H_0 and $\zeta_m < \infty$,

$$\sup\left\{\left|Z_n^*(t) - Z_n(t) + \int_{0^+}^{t^-} s^{-1} Z_n(s)\, ds\right| : 0 \leq t \leq 1\right\} \overset{p}{\to} 0 \quad \text{as } m \to \infty. \tag{3.2.17}$$

Also, if we define $Z^* = \{Z^*(t); t \in [0, 1]\}$ by letting $Z^*(t) = Z(t) - \int_0^t s^{-1} Z(s)\, ds$, $0 \leq t \leq 1$, where Z is a standard Wiener process, then it is easy to verify that Z^* is also a standard Wiener process on $[0, 1]$. As such, by (3.2.15), (3.2.17) and some standard steps (viz. Sen (1982c)), it follows that under H_0, whenever $0 < \zeta_1 \leq m^{-1}\zeta_m < \infty$,

$$Z_n^* \underset{\mathscr{D}}{\Rightarrow} Z, \quad \text{in the } J_1\text{-topology on } D[0, 1]. \tag{3.2.18}$$

Note that (cf. Sen (1960)) the estimator s_n^2 in (3.2.12) is a variant form of the jackknifed estimator, and it follows from Sproule (1974) and Sen (1977b) that under H_0 and the same regularity condition (that $\zeta_m < \infty$),

$$s_n^2 \to \zeta_1 \quad \text{almost surely (a.s.) as } n \to \infty. \tag{3.2.19}$$

As such, by (3.2.13)–(3.2.14), (3.2.18) and (3.2.19), we obtain by some standard arguments that under H_0, D_n^+ (and D_n) have asymptotically the same distribution as $D^+ = \sup\{Z(t) : 0 \leq t \leq 1\}$ (and $D = \sup\{|Z(t)| : 0 \leq t \leq 1\}$). These distributions are given in (2.4.7)–(2.4.8). Basically, $\delta_{n,\alpha}^+ \to \tau_{\alpha/2}$, the upper $50\alpha\%$ point of the standard normal d.f., while, for small α, $\delta_{n,\alpha} \to \tau_{\alpha/4}$. We may remark that instead of the unweighted W_k in (3.2.13)–(3.2.14), we could have taken their standardized versions $(k-m)^{-1/2}W_k/\tilde{s}_k$ (and dropping $(n-m)^{-1/2}$), for $k \geq m+1$. In that case, if in the range, $m < k \leq n$, we replace the lower limit m by ηn, for some $\eta > 0$; then we are also able to use appropriate Wiener process approximation with square root boundaries (over the range $[\eta, 1]$), and critical values for these (for various typical values of η) have been tabulated by DeLong (1981). Typically, η may be chosen as $0.01, 0.05$ or 0.10.

The asymptotic distribution theory of U-statistics in the case on i. non-i.d. r.v.'s have been studied in Hoeffding (1948) with further simplications in Sen (1969). If we define

$$\gamma(F, G) = \int g_1^{(G)}(x)\, dF(x) - \theta(G), \qquad g_1^{(G)}(x) = E_G g(x, X_2, \ldots, X_m),$$
(3.2.20)

$$\bar{F}_k(x) = k^{-1} \sum_{i=1}^{k} F_i(x) \quad \text{for every } k \geq 1,$$
(3.2.21)

and

$$\gamma_n^+ = \max_{m < k \leq n} \left\{ (n-m)^{-1} \sum_{i=m+1}^{k} \gamma(F_i, \bar{F}_{i-1}) \right\},$$

$$\gamma_n = \max_{m < k \leq n} \left\{ (n-m)^{-1} \left| \sum_{i=m+1}^{k} \gamma(F_i, \bar{F}_{i-1}) \right| \right\},$$
(3.2.22)

then, it follows from Sen (1982c) that the test based on D_n^+ (or D_n) is consistent for all alternatives for which $\lim \gamma_n^+ > 0$ (or $\lim \gamma_n > 0$). In particular, if (for the change-point) q/n is away from 0 or 1, the lower limits in (3.2.22) are generally positive under very mild regularity conditions, and hence, the tests are consistent for such change-point models.

For local (contiguous) alternatives, asymptotic power properties of these tests have also been studied in Sen (1982c). To fit such a contiguous model, we assume that the change point τ belongs to $(t_{q_n}, t_{q_n+1}]$, where q_n is $\nearrow$ in n with

$$\lim_{n \to \infty} n^{-1}q_n = \pi : 0 < \pi < 1.$$
(3.2.23)

Further, for every n, $F_1 = \cdots = F_{q_n} = F$ and $F_{q_n+1} = \cdots = F_n$ satisfy

$$n^{1/2}\{F_n(x) - F(x)\} = H_n(x) \to H(x),$$
(3.2.24)

where $H(x)$ does not depend on n and it leads to the satisfaction of the condition of contiguity (of the probability measures under the alternative hypothesis to that under H_0). In particular, we need in this context that $\int (dH(x)/dF(x))^2\, dF(x) < \infty$. Let then

$$\gamma = \int g_1^{(F)}(x)\, dH(x) \quad \left(= \int g_1^{(F)}(x)[dH(x)/dF(x)]\, dF(x) \right),$$
(3.2.25)

$$\mu(t) = \begin{cases} 0, & 0 < t < \pi, \\ \pi\{\log(t/\pi)\}\gamma/\zeta_1^{1/2}, & \pi \leq t \leq 1. \end{cases}$$
(3.2.26)

Under such a contiguous alternative, $\{K_n\}$, $\{Z_n^*\}$, defined after (3.2.16), converges weakly to $Z + \mu$; $\mu = \{\mu(t); t \in [0, 1]\}$, so that the asymptotic power functions of the tests based on D_n^+ and D_n are respectively given by

$$P\{Z(t) + \mu(t) \geqq \delta_\alpha^+, \text{ for some } t \in [0, 1]\} \qquad (3.2.27)$$

and

$$P\{|Z(t) + \mu(t)| \geqq \delta_\alpha, \text{ for some } t \in [0, 1]\}, \qquad (3.2.28)$$

where $\delta_\alpha^+ = \lim_{n \to \infty} \delta_{n,\alpha}^+ \ (= \tau_{\alpha/2})$ and $\delta_\alpha = \lim_{n \to \infty} \delta_{n,\alpha}$.

The case of more than one change-point may also be handled in a similar way. In (3.2.23)–(3.2.24), we denote the d.f.'s by F_{ni}, $i = 1, \ldots, n$, and assume that

$$n^{1/2}\{F_{ni}(x) - F(x)\} \to H_i(x), \qquad i = 1, \ldots, n, \qquad (3.2.29)$$

where the H_i can be one of $r \ (\geqq 1)$ different $H^{(1)}, \ldots, H^{(r)}$ quantities, depending on where i/n belongs (one of the r nonoverlapping intervals in $[0, 1]$). With these adjustments and for the $H^{(j)}$ satisfying the condition of contiguity, the asymptotic power functions will be given by (3.2.27) and (3.2.28) but for more complicated (and segmented) forms for the $\mu(t)$.

Note that for the sequential detection problem, one has to replace the domain of Z_n^* (and Z) by $[0, \infty)$ with their definitions extended, and this can easily be done by using the results of Lindvall (1973) which enable us to extend (3.2.18) on $D[0, \infty)$ under no extra condition. This will provide nice asymptotic expressions for the "false alarm" probabilities (i.e., $P\{N > n \mid q = +\infty\}$) where N is the stopping number. Further, the results in § 3.4 of Sen (1981d) on the embedding of Wiener process for U-statistics allow us to transmit this problem to that of a standard Wiener process on $[0, \infty)$. The recent results of Sen (1984d) may be used for finding the asymptotic expressions for $P\{N > q + r \mid q < \infty\} \ (r \geqq 1)$ in terms of a drifted Wiener process. These would then provide useful information on the performance characteristics of the procedure in the sequential detection problem. If, however, one wants to compute $E\{N \mid q = +\infty\}$ and $E\{(N - q)^+ \mid q < \infty\}$, then stronger modes of convergence (not implied by the weak or strong invariance principles) will be needed. Some other tests for change-points based on U-statistics (not of the recursive type or quasi-sequential in nature) will be briefly discussed in the last section.

3.3. Recursive residual rank tests for change-points. With the same setup as in § 3.1, consider the usual linear model where the d.f. F_i of X_i is given by

$$F_i(x) = F(x - \boldsymbol{\beta}_i' \mathbf{c}_i), \quad x \in (-\infty, \infty), \qquad i = 1, \ldots, n, \qquad (3.3.1)$$

where the regressors $\mathbf{c}_i$ are known $q \ (\geqq 1)$-vectors, the $\boldsymbol{\beta}_i$ are unknown regression parameters (vectors) and F is an unspecified, continuous d.f. One then frames the null hypothesis H_0 of the *constancy of regression relationships over time* i.e., $H_0: \boldsymbol{\beta}_1 = \cdots = \boldsymbol{\beta}_n = \boldsymbol{\beta}$ (unknown), against the composite alternative that

$$\boldsymbol{\beta}_1 = \cdots = \boldsymbol{\beta}_m \neq \boldsymbol{\beta}_{m+1} = \cdots = \boldsymbol{\beta}_n, \quad \text{for some (unknown) } m: 1 \leqq m < n. \qquad (3.3.2)$$

The change-point problem relating to the location model corresponds to the particular case of $q = 1$ and $c_i = 1$, $\forall\, i \geq 1$. For this location model, under H_0, the X_i are i.i.d.r.v. and the *recursive ranking* scheme of Bhattacharya and Frierson (1981) works out very naturally. Let R_{ii} be the rank of X_i among $X_1, \ldots, X_i$, for $i \geq 1$. Then, under H_0, the R_{ii} are independent random variables, where for each i, R_{ii} can assume the values $1, \ldots, i$ with the equal probability $1/i$. Thus, if for every i, we define a set of *scores* $a_i(1), \ldots, a_i(i)$ as in § 2.3, then we may define a recursive rank statistic by

$$L_n^* = \sum_{i=1}^{n} [a_i(R_{ii}) - \bar{a}_i] \quad \text{where } \bar{a}_i = i^{-1} \sum_{j=1}^{i} a_i(j), \quad i \geq 1. \tag{3.3.3}$$

Under H_0, L_n^* is a genuinely distribution-free statistic with mean 0 and variance $A_n^{*2} = \sum_{i=1}^{n} A_i^2$, where $A_i^2 = i^{-1} \sum_{j=1}^{i} \{a_i(j) - \bar{a}_i\}^2$, for $i \geq 1$. If these scores are generated by a score function $\phi = \{\phi(u), 0 < u < 1\}$ where ϕ is expressible as the difference of two nondecreasing, absolutely continuous and square integrable functions [inside $(0, 1)$], then $n^{-1} A_n^{*2} \to A^2 = \int_0^1 \phi^2(u)\, du - (\int_0^1 \phi(u)\, du)^2$. Further, under the same condition, the usual (weak as well as strong) invariance principles (under H_0) hold for the sequence $\{L_n^*\}$, so that we may proceed as in § 3.2 and consider the test statistics $D_n^+ = \max\{L_k^*/A_n^* : 1 \leq k \leq n\}$ and $D_n = \max\{|L_k^*|/A_n^* : 1 \leq k \leq n\}$; here, both D_n^+ and D_n are genuinely distribution-free under H_0, and they have the same limiting (null) distributions as in (2.4.7)–(2.4.8). For local (contiguous) alternatives, the nonnull (asymptotic) distributions of these statistics will also be parallel to (3.2.27)–(3.2.28) where the drift function is segmented-linear, i.e., $\mu(t) = 0$ for $t \leq \pi$ and $\zeta(t - \pi)$ for $\pi < t \leq 1$, where ζ is some constant depending on the local shift under the alternative. This procedure, though very simple, may not work out for the general regression model in (3.3.2) (where the $\mathbf{c}_i$ may not be all equal). The basic reason for this is that when the $\mathbf{c}_i$ are not all the same, even under H_0 ($\boldsymbol{\beta}$ unknown), the X_i fail to be identically distributed, so that the basic independence of the recursive ranks R_{ii}, $i \geq 1$ may not hold, and without this property, the distribution-freeness and other properties of the test statistics also may not follow. To overcome this problem, we consider some recursive residual (signed) rank statistics which provide robust and asymptotically distribution-free tests for the testing problem (3.3.1)–(3.3.2), where the $\mathbf{c}_i$ need not be all equal (or groupwise equal) and $q\ (\geq 1)$ may be arbitrary.

We assume that the d.f. F in (3.3.1) is symmetric about 0. Let $\phi^+ = \{\phi^+(u), 0 < u < 1\}$ be a nonconstant, nondecreasing and square integrable score function where $\phi^+(u) = \phi((1 + u)/2)$, $0 < u < 1$, and ϕ is skew-symmetric i.e., $\phi(u) + \phi(1 - u) = 0$ for every $u \in (0, 1)$. The scores based on ϕ^+, defined as in § 0.0 are denoted by $a_i^+(j)$, $j = 1, \ldots, i$; $i \geq 1$. Assuming that H_0 is true, based on $X_1, \ldots, X_k$, let $\hat{\boldsymbol{\beta}}_k$ be some suitable estimator of $\boldsymbol{\beta}$, for $k \geq 1$. These estimators are quite arbitrary (viz. least squares, R-estimators or M-estimators) and it is assumed that under H_0, for every $\varepsilon > 0$, there exists an integer $k_0\ (\geq 1)$ such that for every $n > k_i$

$$P\left\{ \max_{k_0 \leq k \leq n} (\log k)^{-1} k^{1/2} \|\hat{\boldsymbol{\beta}}_k - \boldsymbol{\beta}\| > 1 \right\} < \varepsilon. \tag{3.3.4}$$

It has been shown in Sen (1983a) that (3.3.4) holds for the various common estimators under quite general regularity conditions. Then, at the kth stage, we define the (recursive) residuals as

$$\hat{X}_{ki} = X_i - \hat{\boldsymbol{\beta}}'_{k-1}\mathbf{c}_i, \qquad i = 1, \ldots, k, \quad \text{for } k = 1, \ldots, n. \tag{3.3.5}$$

(Usually, for $k \leq q$, these residuals are all equal to 0.) Let then $\hat{R}^+_{ki}$ be the rank of $|\hat{X}_{ki}|$ among $|\hat{X}_{k1}|, \ldots, |\hat{X}_{kk}|$, for $i = 1, \ldots, k$; $k \geq 1$. We consider the residual signed rank scores

$$\hat{u}_k = \text{sign}\,(\hat{X}_{kk})a_k^+(\hat{R}^+_{kk}), \qquad k = q+1, \ldots, n, \tag{3.3.6}$$

and, conventionally, we let $\hat{u}_k = 0$ for $k \leq q$. Then, the CUSUMs for these residual rank scores are

$$\hat{U}_k = \sum_{i \leq k} \hat{u}_i \quad \text{for } k = 1, \ldots, n. \tag{3.3.7}$$

Finally, we define (the score-variance) A^2 as in earlier and let

$$\hat{D}_n^+ = (n-q)^{-1/2}A^{-1}\Big\{\max_{k \leq n} \hat{U}_k\Big\} \quad \text{and} \quad \hat{D}_n = (n-q)^{-1/2}A^{-1}\Big\{\max_{k \leq n} |\hat{U}_k|\Big\}. \tag{3.3.8}$$

Then, the recursive residual rank tests for the constancy of regression relationship may be based on $\hat{D}_n^+$ and $\hat{D}_n$. Note that unlike the location model, here the use of $\hat{D}_n^+$ is advocated only where, under the alternative, the $\boldsymbol{\beta}'_k\mathbf{c}_k$ are monotone. Since this may not generally be the case, the two-sided test based on $\hat{D}_n$ is more generally recommended.

The distribution theory of D_n^+ and D_n has been studied in Sen (1983a) by studying the affinity of these statistics to their counterparts when $\boldsymbol{\beta}$ is known. Under H_0, let us define $Y_i = X_i - \boldsymbol{\beta}'\mathbf{c}_i$, $i = 1, \ldots, n$, so that the Y_i are i.i.d.r.v.'s with the d.f. F, symmetric about 0. Let R^+_{ki} be the rank of $|Y_i|$ among $|Y_1|, \ldots, |Y_k|$, for $i = 1, \ldots, k$; $k = 1, \ldots, n$, and let

$$u_k = \text{sign}\,(Y_k)a_k^+(R^+_{kk}) \quad \text{and} \quad U_k = \sum_{i \leq k} u_i \quad \text{for } k = 1, \ldots, n. \tag{3.3.9}$$

Also, we define D_n^+ and D_n as in (3.3.8) with the $\hat{U}_k$ being replaced by U_k. Note that under H_0, (i) sign Y_i and R^+_{ii} are mutually independent, (ii) sign Y_i assumes the values ± 1 with equal probability $\frac{1}{2}$ while R^+_{ii} assumes the values $1, \ldots, i$, with the common probability $1/i$, and (iii) for different i, (sign Y_i, R^+_{ii}) (and hence, the u_i) are stochastically independent. Hence, the usual (weak as well as strong) invariance principles (under H_0) hold for the U_k under the standard regularity conditions, and under H_0, the asymptotic distributions of D_n^+ and D_n are again given by (2.4.7) and (2.4.8). With these results in hand, we may remark that for the asymptotic distribution-freeness of $\hat{D}_n^+$ and $\hat{D}_n$, all we need to show is that under H_0 and suitable regularity conditions,

$$\max\Big\{n^{-1/2}\Big|\sum_{i \leq k}(\hat{u}_i - u_i)\Big| : k \leq n\Big\} \xrightarrow{P} 0 \quad \text{as } n \to \infty. \tag{3.3.10}$$

The regularity conditions under which (3.2.10) was established in Sen (1983a) are the following:

(i) The d.f. F has bounded and continuous first and second order derivatives (f and f', respectively), and (finite Fisher information)

$$I(f) = \int_{-\infty}^{\infty} \{f'(x)/f(x)\}^2 \, dF(x) < \infty. \tag{3.3.11}$$

(ii) The score function ϕ has first and second order derivatives $\phi^{(1)}$ and $\phi^{(2)}$, and there exist a generic constant $K \, (<\infty)$ and a $\delta \, (<\frac{1}{6})$, such that

$$|\phi^{(r)}(u)| \leq K[u(1-u)]^{-r-\delta}, \qquad 0 < u < 1, \quad r = 0, 1, 2. \tag{3.3.12}$$

The condition on δ may be slightly relaxed under a counter condition on f.

(iii) There exists a positive definite (p.d.) and finite matrix $\mathbf{C}_0$, such that

$$n^{-1}\mathbf{C}_n = n^{-1} \sum_{i=1}^{n} \mathbf{c}_i \mathbf{c}_i' \to \mathbf{C}_0 \quad \text{as } n \to \infty,$$

$$\max \{\mathbf{c}_k' \mathbf{C}_0^{-1} \mathbf{c}_k : 1 \leq k \leq n\} = O((\log n)^2). \tag{3.3.13}$$

The second condition on the $\mathbf{c}_i$ may be replaced by the usual Noether condition (i.e., $\max \{n^{-1/2} \|\mathbf{c}_i\| : 1 \leq i \leq n\} \to 0$, as $n \to \infty$) when $\phi^{(2)}$ is bounded.

For the proof of (3.3.10) under the regularity conditions (i), (ii) and (iii) in (3.3.11)–(3.3.13), we may refer to Sen (1983a); it is quite elaborate and is based on the intricate properties of the empirical distribution functions. In this context, we may note that $k^{-1}R_{kk}^+ = \hat{H}_k(|\hat{X}_{kk}|)$, where $\hat{H}_k(y) = k^{-1}\sum_{i=1}^{k} I(y \geq |\hat{X}_{ki}|)$, $y \in R^+$ is the empirical d.f. of the absolute recursive residuals at the kth stage, for $k \geq 1$. Further, the scores $a_k^+(\hat{R}_{kk}^+)$ can safely be approximated by $\phi((k+1)^{-1}k\hat{H}_k \, (|\hat{X}_{kk}|))$, $k \geq 1$. If $H_k(y) = k^{-1}\sum_{i=1}^{k} I(y \geq |Y_i|)$, $y \in R^+$ stands for the empirical d.f. of the true residuals (absolute values) at the kth stage (under H_0), $k \geq 1$, then, essentially, using (3.3.4), (3.3.13), some invariance principles for these empirical d.f.'s and the Taylor expansion on the $\phi(\cdot)$, one approximates $\hat{H}_k(|\hat{X}_{kk}|)$ by $H_k(|Y_k|)$ and employs the same in obtaining a similar approximation for $a_k^+(\hat{R}_{kk}^+)$ by $a_k^+(R_{kk}^+)$, $k \geq 1$. For the negligibility of the remainder terms, (3.3.12) plays the vital role. In passing, we may remark that (3.3.12) holds for all the commonly adopted scores in practice (viz. Wilcoxon scores, $\phi(u) = 2u - 1$ and normal scores $\phi(u) = \Phi^{-1}(u)$, $0 < u < 1$, the inverse of the standard normal d.f.). Also, (3.3.11) encompasses a broad class of density functions, while (3.3.13) holds generally for the replicated models where one may even replace $O(\log n^2)$ by $O(1)$.

Let us next consider the nonnull distribution theory of these test statistics. The consistency of the tests follows as in § 3.2, and hence, we shall again confine ourselves to some local (contiguous) alternatives where these asymptotic distributions are nondegenerate and provide meaningful power properties of these tests. As in (3.2.23)–(3.2.24), we assume here that the change-point

τ $(= \tau_n)$ satisfies the condition that as $n \to \infty$,

$$t_{m(n)} \leq \tau_n < t_{m(n)+1} \quad \text{where } n^{-1}m(n) \to \pi : 0 < \pi < 1, \qquad (3.3.14)$$

$$\boldsymbol{\beta}_1 = \cdots = \boldsymbol{\beta}_{m(n)} = \boldsymbol{\beta}_{m(n)+1} - n^{-1/2}\boldsymbol{\lambda}, \qquad \boldsymbol{\beta}_{m(n)+1} = \cdots = \boldsymbol{\beta}_n, \qquad (3.3.15)$$

where $\boldsymbol{\lambda}$ $(\neq \mathbf{0})$ is fixed. Note that under (3.3.11) through (3.3.15), whenever $\bar{\mathbf{c}}_m = m^{-1}\sum_{i=1}^{m} \mathbf{c}_i \to \bar{\mathbf{c}}$, as $m \to \infty$, the contiguity of the probability measure under the alternative to that under H_0 follows by reference to the standard arguments given in Hájek and Šidák (1967, Chap. VI). Thus the asymptotic equivalence of $\hat{D}_n$ and D_n (or $\hat{D}_n^+$ and D_n^+) under H_0 and this contiguity insure the same under the alternatives specified by (3.3.14)–(3.3.15). For D_n^+ and D_n, based on the partial sums U_k in (3.5.9), we may again appeal to the contiguity and obtain an invariance principle under such an alternative: This relates to a drifted Brownian motion with the drift function $\mu = \{\mu(t), t \in [0, 1]\}$ given by

$$\mu(t) = \begin{cases} 0, & 0 \leq t \leq \pi, \\ A^{-1}\gamma(\phi, F) \cdot \boldsymbol{\lambda}'\bar{\mathbf{c}}\pi \log (t/\pi), & \pi < t \leq 1, \end{cases} \qquad (3.3.16)$$

where

$$\gamma(\phi, F) = \int_{-\infty}^{\infty} (d/dx)\phi(F(x)) \, dF(x)$$

$$= \int_{-\infty}^{\infty} \phi(F(x))\{-f'(x)/f(x)\} \, dF(x) \quad (>0). \qquad (3.3.17)$$

At this stage, we record a small error in Sen (1983a) where due to a computational slip, the drift function in (3.3.16) was written as $(t - \pi) \cdot A^{-1}\gamma(\phi, F)\boldsymbol{\lambda}'\bar{\mathbf{c}}$ (i.e., instead of the logarithm drift for $t > \pi$, a linear drift was reported). As such, for the asymptotic power functions of the tests based on $\hat{D}_n^+$ and $\hat{D}_n$, we have the same formulae as in (3.3.27) and (3.2.28) with the drift $\mu(t)$ given by (3.3.16). Note that the drift in (3.3.16) and in (3.2.26) are proportional to each other. Because of this in some special cases (viz. change in locations), where both the tests in §§ 3.2 and 3.3 are applicable, one may study the asymptotic relative efficiency (ARE) result by the proportionality factor $\gamma^2 A^2/\{\zeta_1\gamma^2(\phi, F)\}$, which agrees with the conventional Pitman-ARE result. Moreover, within the class of (aligned) recursive residual rank tests for change points (based on different score functions), the drifts would be proportional to each other, and hence, again the usual measure of Pitman-ARE of (nonsequential) rank tests can be adopted in this context too. For more details, we may refer to Sen (1983a).

As in (3.2.29), one may also consider here the case of more than one change point (in a local (contiguous) alternative setup), and the asymptotic power function will involve segmented logarithmic functions in the drift function. Further, the sequential detection problem based on these recursive residual signed rank statistics can also be treated as in § 3.2. In this context, some almost sure invariance principles for the recursive residual rank statistics are needed and can be worked out in the same manner.

3.4. Recursive residual M-tests. We consider the same setup as in (3.3.1) and (3.3.2). Instead of using recursive rank statistics, we will incorporate here some M-statistics for constructing some robust tests for change-points. Let $\psi = \{\psi(x), \; x \in (-\infty, \infty)\}$ be a nondecreasing, continuous and skew-symmetric *score function* and for every $\mathbf{B} = (b_1, \ldots, b_q)' \in E^q$, define

$$\mathbf{M}_n(\mathbf{b}) = \sum_{i=1}^{n} \mathbf{c}_i \psi(X_i - \mathbf{b}'\mathbf{c}_i), \qquad n \geq 1, \tag{3.4.1}$$

where the regression vectors $\mathbf{c}_i$ are known and associated with the model (3.3.1). Note that $\int \psi(x) \, dF(x) = 0$ (by virtue of the skew-symmetry of ψ and symmetry of F), so that $\mathbf{M}_n(\boldsymbol{\beta})$ is centered around $\mathbf{0}$ when $\boldsymbol{\beta}_1 = \cdots = \boldsymbol{\beta}_n = \boldsymbol{\beta}$. The M-estimator of $\boldsymbol{\beta}$, based on the score function ψ and the observations $X_1, \ldots, X_n$, is then defined by the solution of

$$\mathbf{M}_n(\hat{\boldsymbol{\beta}}_n) = 0, \qquad \hat{\boldsymbol{\beta}}_n = (\beta_{n,1}, \ldots, \beta_{n,q})'. \tag{3.4.2}$$

In the same manner as in (3.4.2), for every $k \geq q$, we define the M-estimator $\hat{\boldsymbol{\beta}}_k$, and consider the recursive residuals

$$\hat{X}_{ki} = X_i - \hat{\boldsymbol{\beta}}_{k-1}' \mathbf{c}_i, \quad \text{for } i = 1, \ldots, k, \quad q < k \leq n; \tag{3.4.3}$$

for $k \leq q$, we take, conventionally, $\hat{X}_{ki} = 0$, $i = 1, \ldots, k$. Further, we define

$$s_k^2 = k^{-1} \sum_{i=1}^{k} \psi^2(\hat{X}_{ki}), \quad k > q, \qquad s_k^2 = 0 \quad \text{for } k \leq q. \tag{3.4.4}$$

Then, the recursive M-statistics are defined by

$$M_k^* = \sum_{i=1}^{k} \psi(\hat{X}_{ii}) \quad \text{for } k = 0, 1, \ldots, n. \tag{3.4.5}$$

Analogous to (3.3.7)–(3.3.8), we define here

$$\hat{D}_n^+ = (n-q)^{-1/2}\left\{\max_{k \leq n} M_2^*/\tilde{s}_k\right\} \quad \text{and} \quad \hat{D}_n = (n-q)^{-1/2}\left\{\max_{k \leq n} |M_k^*|/\tilde{s}_k\right\}, \tag{3.4.6}$$

where $\tilde{s}_k = \max\{s_k, k^{-1/2}\}$, $k \geq 1$. Then the test for the change-point problem may be based on $\hat{D}_n^+$ (for one-sided alternatives) and $\hat{D}_n$ (for two-sided ones). We may note that the observations X_i are taken sequentially at the ordered time points t_i and $M_k^*/\tilde{s}_k$ is based only on the set $\hat{X}_{ki}$, $i \leq k$, for $k \geq 1$. Thus, the test for $H_0: \boldsymbol{\beta}_1 = \cdots = \boldsymbol{\beta}_n = \boldsymbol{\beta}$ (unknown) against the change-point alternative in (3.3.2) may be carried out in a quasi-sequential manner. The basic problem is to determine the critical values of D_n^+ and D_n. Operationally, the test procedure is very similar to the one in § 3.2 (based on recursive U-statistics) and the discussion following (3.2.14) also pertains here. Thus, we need to study some invariance principles for the recursive M-statistics $\{M_k^*\}$, which would then provide the desired asymptotic approximations.

For this study, we assume further that ψ has a bounded derivative inside

$(-k, k)$ (where $0 < k < \infty$) and $\psi(x) = \psi(k)$ sign x for $|x| \geq k$. This boundedness condition is usually advocated to induce robustness against outliers. Concerning the $\mathbf{c}_i$, we assume that the same regularity conditions as in § 3.3 hold. Let us define then

$$\gamma = \int_{-\infty}^{\infty} \psi'(x)\, dF(x) = \int_{-k}^{k} \psi'(x)\, dF(x) \quad (>0), \qquad (3.4.7)$$

$$\sigma_\psi^2 = \int_{-\infty}^{\infty} \psi^2(x)\, dF(x) \quad \text{and} \quad \nu_\psi^2 = \sigma_\psi^2 / \gamma^2. \qquad (3.4.8)$$

Note that by assumption $\nu_\psi < \infty$, while we assume further that $\nu_\psi > 0$. We may note further that by virtue of the assumed boundedness condition on ψ, for every sequence $\{k_n\}$ of positive integers, such that $k_n \uparrow \infty$ but $n^{-1}k_n^2 \downarrow 0$ as $n \to \infty$, we have $\max\{n^{-1/2}|M_k^*| : k \leq k_n\} > 0$, with probability 1. Also, exploiting the asymptotic linearity of $\mathbf{M}_k(\mathbf{b})$ in $\mathbf{b}$ (in the neighbourhood of $\boldsymbol{\beta}$) (viz., Jurečková and Sen (1984)), it has been shown (by Sen (1984e)) that under the assumed regularity conditions, as $n \to \infty$, when H_0 holds

$$\max_{k_n \leq k \leq n} k^{9/10} \|(\hat{\boldsymbol{\beta}}_k - \boldsymbol{\beta}) - \gamma^{-1}\mathbf{M}_k(\boldsymbol{\beta})\mathbf{C}_k^{-1}\| = O_p((\log\log n)^{1/2}), \qquad (3.4.9)$$

where the $\mathbf{C}_k$ are defined as in (3.3.13). Further, note that under H_0,

$$k^{-1} \sum_{i=1}^{k} \psi^2(X_i - \boldsymbol{\beta}'\mathbf{c}_i) = s_k^{02} \to \sigma_\psi^2 \quad \text{as } k \to \infty, \qquad (3.4.10)$$

while, by virtue of the implied Lipschitz condition on ψ, we have

$$\max_{k_n^* \leq k \leq n} |s_k^2 - s_k^{02}| = \max_{k_n^* \leq k \leq n} \left| k^{-1} \sum_{i=1}^{k} \{\psi^2(X_i - \hat{\boldsymbol{\beta}}_{k-1}'\mathbf{c}_i) - \psi^2(X_i - '\mathbf{c}_i)\} \right|$$

$$= \max_{k_n^* \leq k \leq n} \left\{ O(\|\hat{\boldsymbol{\beta}}_{k-1} - \boldsymbol{\beta}\|) \cdot \max_{1 \leq i \leq k} \|\mathbf{c}_i\| \sup_x |\psi(x)| \right\}$$

$$= \max_{k_n^* \leq k \leq n} \{ O_p(k^{-9/10} \log\log n) \cdot O((\log k)^2) \cdot O(1) \}$$

$$= o_p(1), \quad \text{for every } k_n^* : (\log n)^3 / k_n^* \to 0, \qquad (3.4.11)$$

as $n \to \infty$, and in this context, we have made use of (3.4.9) in the third step. Consequently, by (3.4.10) and (3.4.11), we conclude that under H_0, for every $\{k_n^*\}$ increasing at a faster rate than $(\log n)^3$,

$$\max_{k_n^* \leq k \leq n} |s_k^2 - \sigma_\psi^2| \to 0 \quad \text{in probability,} \quad \text{as } n \to \infty. \qquad (3.4.12)$$

Let us denote by $\tilde{E}_k$ the conditional expectation given $X_i, i \leq k$, and let $Z_k = \psi(\hat{X}_{kk}) - \psi(X_k - \boldsymbol{\beta}'\mathbf{c}_k) - \tilde{E}_{k-1}\psi(\hat{X}_{kk})$, $k \geq 1$. Note that the Z_k form a martingale sequence, and $E(Z_k^2) \to 0$, as $k \to \infty$. Consequently, using the weak invariance principles for martingales, we conclude that (for details, we refer to

Sen (1984b))

$$\max_{k \leq n} \left\{ n^{-1/2} \left| \sum_{i \leq k} Z_i \right| \right\} \to 0 \quad \text{in probability,} \quad \text{as } n \to \infty, \qquad (3.4.13)$$

$$\max_{k_n \leq k \leq n} \left| n^{-1/2} \sum_{i \leq k} \{ \bar{E}_{i-1} \psi(\hat{X}_{ii}) + \gamma(\hat{\boldsymbol{\beta}}_{i-1} - \boldsymbol{\beta})' \mathbf{c}_i \} \right| = o_p(1), \qquad (3.4.14)$$

so that, under H_0, as $n \to \infty$,

$$\max_{k_n \leq k \leq n} \left| n^{-1/2} \sum_{i \leq k} \{ \psi(\hat{X}_{ii}) - \psi(X_i - \boldsymbol{\beta}' \mathbf{c}_i) + \gamma(\hat{\boldsymbol{\beta}}_{i-1} - \boldsymbol{\beta})' \mathbf{c}_i \} \right| \xrightarrow{P} 0. \qquad (3.4.15)$$

By using (3.4.9) and (3.4.15), we obtain that for all $k : k_n \leq k \leq n$,

$$n^{-1/2} \sum_{i \leq k} \{ \psi(X_i - \boldsymbol{\beta}' \mathbf{c}_i) - \gamma(\hat{\boldsymbol{\beta}}_{i-1} - \boldsymbol{\beta})' \mathbf{c}_i \} = n^{-1/2} \sum_{i \leq k} \left(\sum_{j \leq i} a_{ij} \psi(X_j - \boldsymbol{\beta}' \mathbf{c}_j) \right) + \xi_{nk}$$

$$(3.4.16)$$

where $a_{ii} = 1$, $a_{ij} = -\mathbf{c}_j' \mathbf{C}_{i-1}^{-1} \mathbf{c}_i$ for $j < i$ and $a_{ij} = 0$, $j > i$, and where

$$\max \{ |\xi_{nk}| : k_n \leq k \leq n \} = O_p(n^{-2/5} \log \log n), \qquad (3.4.17)$$

where in (3.4.16)–(3.4.17), we choose $k_n = [\varepsilon n^{-1/2}]$, for some arbitrarily small $\varepsilon > 0$. Since the $\psi(X_i - \boldsymbol{\beta}' \mathbf{c}_i)$ are i.i.d.r.v. with mean 0, a finite positive variance σ_ψ^2 and are bounded (so that higher order moments exist), invariance principles for recursive residuals studied in Sen (1982a) are directly applicable to them. Hence, by (3.4.15), (3.4.16) and Sen (1982a, Thm. 1), we conclude that under H_0 and the assumed regularity conditions, for the partial sum process in (3.4.5), the weak invariance principle holds. Hence, under H_0 using (3.4.12) and this weak invariance principle, we conclude that for D_n^+ and D_n in (3.4.6), the limiting distributions in (2.4.7) and (2.4.8) hold. With this asymptotic distribution theory, we may proceed as in § 3.2 or 3.3 and carry out the change-point test based on the recursive M-statistics in (3.4.5). The tests in this section are mostly recommended on the ground of robustness against local departures from some assumed model (related to F), while the tests in the earlier sections on global robustness where F is not assumed to be of some particular type. In the restricted domain where F is nearly known, these M-tests may have better performance than the other ones; however, outside such a restricted domain, the others may have better performance.

As in § 3.3, we now discuss the nonnull distribution theory of these M-tests. For the change-point τ $(= \tau_n)$, we assume that (3.3.14) and (3.3.15) hold. Further, for the symmetric d.f. F, we assume that we have a finite and positive Fisher information $I(f)$. Then, the contiguity of the probability measures under the sequence of alternatives in (3.3.14)–(3.3.15) to that under H_0 follows as in § 3.3. The weak invariance principle under H_0, studied earlier, and this contiguity enable us to establish the weak invariance principle under the alternatives as well. We need to study the drift function only, and incorporate that in the weak convergence result. If we denote by $\{K_n\}$ the sequence of (contiguous) alternatives in (3.3.14)–(3.3.15), then note that for $m(n) \geqq k$,

$$E\left\{ n^{-1/2} \sum_{i \leq k} \sum_{j \leq i} \psi(X_j - \boldsymbol{\beta}' \mathbf{c}_j) \mid K_n \right\} = 0,$$

which for $k > m(n) = m$,

$$E\left\{ n^{-1/2} \sum_{i \leq k} \sum_{j \leq i} \psi(X_j - \boldsymbol{\beta}' \mathbf{c}_j) \mid K_n \right\} = n^{-1} \sum_{m < i \leq k} \sum_{m < j \leq i} a_{ij} \gamma \boldsymbol{\lambda}' \mathbf{c}_j + o(1)$$

$$= \gamma \boldsymbol{\lambda}' \bar{\mathbf{c}} \cdot \{(m/n) \log (k/m)\} + o(1),$$

$$\tag{3.4.18}$$

where in the above derivation, as in § 3.3, we assume that $\bar{\mathbf{c}}_k \to \bar{\mathbf{c}}$ as $k \to \infty$, $|\bar{\mathbf{c}}| < \infty$, and we have made use of the definitions of the a_{ij} (as in after (3.4.16)) along with the fact that $\mathbf{c}_i \mathbf{c}_i' = \mathbf{C}_i - \mathbf{C}_{i-1}$, for every $i \geqq 1$, where $k^{-1} \mathbf{C}_k$ converges to a p.d. limit $\mathbf{C}_0$, as $k \to \infty$. For some of these details, we may again refer to Sen (1984e). Hence, if we define $\omega = \{\omega(t), t \in [0, 1]\}$ by letting

$$\omega(t) = \begin{cases} 0, & 0 \leq t \leq \pi, \\ \sigma_\psi^{-1} \gamma \boldsymbol{\lambda}' \bar{\mathbf{c}} \{\pi \log (t/\pi)\}, & \pi < t \leq 1, \end{cases} \tag{3.4.19}$$

where π is defined by (3.3.14), then, under $\{K_n\}$, for the recursive M-statistics in (3.4.5), we have the weak convergence to a drifted Wiener process with the drift function ω given by (3.4.19). As such, for the asymptotic distribution of D_n^+ or D_n in (3.4.6), we have the same result as in (3.2.27) or (3.2.28), where the drift μ has to be replaced by ω. In this sense, for all the recursive residual tests based on U-statistics, signed-rank statistics and M-statistics, we have the same form of the asymptotic null distributions and similar forms of the nonnull distributions, where the drifts (though nonlinear) are proportional to each other. If we compare these nonnull distributions, we come up with the usual measure of the asymptotic relative efficiency (classically known as the Pitman efficiency), though here we do not have the asymptotic distribution of the type of a normal or chi-square distribution. The discussions following (3.2.29) on the situation involving more than one change-point and for the sequential detection problem remain true for such recursive M-tests as well.

3.5. Some general remarks. In §§ 3.2, 3.3 and 3.4, we have mainly concentrated on the quasi-sequential procedures. For the change-point problem, we may also have the situation where all the n observations (along with the time points $t_1, \ldots, t_n$) are given, and we may like to know whether a change-point has occurred during (t_1, t_n). In such a case, there seems to be no real need for a sequential procedure, and there are other procedures which work out quite well. In the simple nonparametric problem, described in § 3.1, against shift alternatives (i.e., (3.3.1) with the $\mathbf{c}_i = 1$ and $\boldsymbol{\beta}_i = \theta_i$ standing for the location parameters), some genuinely distribution-free tests for the change-point problem have been considered by a host of workers. Page (1955) considered a test based on the cumulative sums of the sign $(X_i - \theta_0)$ when the initial median θ_0 is specified; in the same case, Bhattacharyya and Johnson (1968) considered a general class of locally most powerful rank tests based on appropriate signed rank statistics. These tests are distribution-free under H_0, while the recursive residual signed tests in § 3.3 are only asymptotically distribution-free. When the initial median θ_0 is unknown, Bhattacharyya and Johnson (1968) considered some locally most powerful invariant tests based on simple linear rank statistics. Again, for the simple shift model, their statistics are genuinely distribution-free (under H_0), whereas, their recursive counterparts are not so. Nevertheless, for these genuinely distribution-free tests, one needs to assume that under H_0 all the n observations are i.i.d., while, in § 3.3 and 3.4, we have a more general model incorporating regression relationships, and under H_0, the constancy of these regression relationships does not necessarily ensure the i.i.d. structure of the observable r.v.'s. Sen and Srivastava (1975) and Pettitt (1979) have considered variant forms of the two-sample Wilcoxon–Mann–Whitney statistics and proposed some maximal statistics similar in structure to D_n^+ and D_n. Like the others, these are also restricted to the case where under H_0, all the n r.v.'s are assumed to be i.i.d., and this is not the case with our §§ 3.3 and 3.4. In the case of i.i.d. structure (under H_0), the tests by Bhattacharya and Frierson (1981) and Lombard (1981), (1983) deserve mention too. If the actual testing problem related to a null hypothesis of i.i.d.r.v.'s against an appropriate (viz. shift) alternative, some of these genuinely distribution-free tests (worked out in a nonsequential setup) may compare quite favorably with the alternative ones studied in §§ 3.3 and 3.4 (when we do not confront a quasi-sequential problem from the operational point of view). However, many of these tests would not be valid for the change-point model relating to the constancy of the regression relationship over time, and hence, their genuine distribution-freeness for a restricted null hypothesis should not be over emphasized. For some other asymptotically distribution-free tests for the constancy of regression relationships over time we may refer to Sen (1977b), (1980c), (1982b). Also, for the tests considered by Sen and Srivastava (1975) and Pettitt (1979), the asymptotic theory rests on an invariance principle for a general class of linear rank statistics, studied in Sen (1978b, § 6); Schechtman and Wolfe (1981) have presented some numerical and simulation studies relating to these. More work on this line would bridge the gap between the asymptotic theory and moderate sample approximations.

From the mathematical point of view, the invariance principles in §§ 3.2, 3.3 and 3.4, though they provide the desired asymptotic theory, are not by themselves sufficient to prescribe suitable rates of convergence for these asymptotic results to hold. There seems to be a pressing need to develop the Edgeworth type expansions for these asymptotic distributional results to yield suitable rates of convergence. More numerical studies in this context would also be very desirable.

In earlier sections of this chapter, we have referred to the sequential detection problem. Though the methodology for this problem runs parallel to that in the change-point problem, from the mathematical point of view, we may require a somewhat more complicated treatment to cover the infinite sequence which may arise in this context. Specifically, for the change-point problem, we need the weak convergence results on the (recursive) statistics and the convergence of their estimated variances. In a sequential detection problem, either a strong invariance principle has to be worked out, or the weak convergence result has to be extended to the half-line $R^+ = [0, \infty)$. For recursive residual rank statistics and recursive M-statistics, this may require stronger modes of convergence for the related R- and M-estimators as well as the aligned statistics. For R- and M-estimators, such stronger modes of convergence have been studied through the "asymptotic almost sure linearity results", and we may refer to Sen and Ghosh (1971), Ghosh and Sen (1972), Jurečková and Sen (1981a, b), (1984) and Sen (1984e), among others. Fortunately, these results are directly adaptable in the current context.

In comparing two tests for the change-point model based on recursive statistics, we have seen in §§ 3.2, 2.2 and 3.4 that the classical Pitman-ARE may be adopted. On the other hand, if one wants to compare a recursive and a nonrecursive test for the same problem, there may be some technical problems. First, the two null distributions may not be the same (even asymptotically). Secondly, even if we are able to have same or comparable null distributions, for the alternative hypotheses (even in the contiguous case), the two nonnull distributions may not have proportional drifts. For example, in the recursive test case, in earlier sections we have seen that typically we have a segmented logarithmic drift (see (3.4.19) for example), while for a nonrecursive test, we may have a segmented linear drift. For such a situation, the conventional Pitman-ARE may not be validly applicable, and we may be forced to study the relative performance by having extensive numerical or simulation studies. The ARE may as well depend a lot on the value of π, the proportion of observations following which a change-point occurs. A meaningful definition of a single numerical measure of the ARE in such a case remains to be worked out.

We have not addressed the problem of estimating the change-point τ_n. In the simplest model (of equality of the F_i under H_0 against a shift at an unknown time point τ_n), one may consider the first exit time (beyond the critical levels) for the CUSUMs (based on the recursive or nonrecursive statistics), and treat this as an estimator of τ_n. Such an estimator may have some undesirable properties. More complicated parametric estimates are available in the literature. This estimation problem (of τ_n) in a nonparametric (or robust) setup deserves a more systematic investigation. In this context, the work of Bhattacharya and Brockwell (1976) provides a basic coverage of the asymptotic theory, and more work is encouraged. The study of the (second order) asymptotic optimality properties of such estimators constitutes an area of high level mathematical research having good statistical prospects too.

CHAPTER 4

Nonparametrics of Sequential Estimation

4.1. Introduction. Based on well defined stopping rules, nonparametric sequential procedures are adaptable for both the *point* and *interval estimation* problems. In the context of *minimum risk* (point) estimation of a parameter (corresponding to a *loss function* incorporating *cost of sampling*), the *optimal* sample size depends, in general, on the underlying distribution (which is either of unknown form or involves some unknown parameter(s) associated with a specified form), and hence, no *fixed sample size* procedure may be optimal simultaneously for all distributions belonging to a family (or parameter(s) belonging to a set). A similar situation arises in constructing a *bounded width confidence interval* for a parameter (having a prescribed *coverage probability*) when the underlying distribution is either of unknown form or it involves some unknown parameter(s). For both of these problems, sequential procedures work out well. These were developed first for some specific parametric models. In the course of these developments, nonparametric procedures emerged in some meaningful and well defined *asymptotic setups*: Optimal nonparametric procedures exist under less stringent regularity conditions and they cover a wider class of underlying distributions. A closer look at this nonparametric theory reveals that some *invariance principles,* (strong) *consistency* and certain *uniform integrability* results for a broad class of nonparametric statistics provide the basic mathematical tools in these developments. Accordingly, in this chapter, we shall discuss these mathematical results and focus on their fruitful role in the sequential estimation problems.

The (asymptotically) minimum risk sequential point estimation problem in a nonparametric setup is considered in § 4.2. The basic mathematical results pertaining to these procedures are then presented in § 4.3. Generalized Behrens–Fisher (two-sample) models and some multiparameter problems are treated briefly in § 4.4. Sequential confidence sets are considered in § 4.5. The last section deals with some general remarks on all these sequential procedures.

4.2. Asymptotically minimum risk sequential point estimation. Let $\{X_i; i \geq 1\}$ be a sequence of i.i.d.r.v. with a d.f. F (belonging to a family $\mathscr{F}$), and let $\theta = \theta(F)$ be a parameter of interest. Based on a sample $X_1, \ldots, X_n$ of size n, let $T_n (= T(X_1, \ldots, X_n))$ be an estimator of θ, and consider the *loss* incurred due to estimating θ by T_n as

$$L_n(c) = g(|T_n - \theta|) + cn \qquad (c > 0) \qquad (4.2.1)$$

where $g(\cdot)$ is a suitable nondecreasing and nonnegative function on R^+ $(= [0, \infty))$ with $g(0) = 0$ and c denotes the *cost* per unit sample. At this stage, we

45

assume that there exists a positive integer n_0 ($<\infty$), such that $\rho_n(F) = E_F g(|T_n - \theta|)$ exists for every $n \geq n_0$ and $\rho_n(F)$ is $\searrow$ in n with $\lim_{n\to\infty} \rho_n(F) = 0$. Then, the *risk* in estimating θ by T_n is given by

$$\lambda_n(c; F) = E_F L_n(c) = \rho_n(F) + cn, \qquad n \geq n_0, \qquad (4.2.2)$$

where $\rho_n(F)$ is $\searrow$ while cn is $\uparrow$ in n, so that there exists an *optimal* n^0 ($\geq n_0$), $n^0 = n^0(c, F)$, such that $\lambda_{n^0}(c; F) = \inf\{\lambda_n(c; F): n \geq n_0\}$; in case of multiple solutions, one may take the smallest value. However, it is quite clear that $n^0(c, F)$ depends, in general, on the unknown F through $\rho_n(F)$, so that no fixed sample size n may satisfy the above solution for all $F \in \mathcal{F}$. Our basic goal is to find out the minimum risk estimator (MRE) of θ when the risk function is given by (4.2.2) and when F ($\in \mathcal{F}$) is not totally specified. To motivate the sequential procedure, we take recourse to an *asymptotic setup*, where in (4.2.1) we allow c (>0) to converge to 0.

We assume that there exist positive q ($0 < q \leq 1$) and $\rho^*(F)$, such that

$$n^q \rho_n(F) \to \rho^*(F): 0 < \rho^*(F) < \infty, \qquad F \in \mathcal{F}, \quad \text{as } n \to \infty. \qquad (4.2.3)$$

In passing, we may remark that usually one chooses $g(y) = y$ or y^2 and the corresponding q is then typically $\frac{1}{2}$ or 1. Note that by (4.2.2) and (4.2.3), as $c \downarrow 0$,

$$n_c^0 = n^0(c, F) \sim \{c^{-1} q \rho^*(F)\}^{(1+q)^{-1}}, \qquad (4.2.4)$$

$$\lambda_F^0(c) = \lambda_{N_c}(c, F) \sim \{c^q(q + q^{-q})\rho^*(F)\}^{(1+q)^{-1}}, \qquad (4.2.5)$$

where $a_c \sim b_c$ means that $\lim_{c\downarrow 0}(a_c/b_c) = 1$. Thus, $\lambda_F^0(c)$ is the *asymptotic minimum risk* (AMR) and, as $c \downarrow 0$, $\lambda_F^0(c) \downarrow 0$.

Next, we assume that there exists a sequence $\{d_n = d(X_1, \ldots, X_n)\}$ of (strongly) consistent estimators of $\rho^*(F), F \in \mathcal{F}, n \geq n_0$. Then, motivated by (4.2.4) and the stochastic convergence of d_n to $\rho^*(F)$, for every c (>0), we may define a *stopping variable* N_c by letting

$$N_c = \inf\{n \geq n_0: n^{q+1} \geq c^{-1} q(d_n + n^{-h})\}. \qquad (4.2.6)$$

where h (>0) is a suitable positive constant, mainly introduced to protect N_c from being unduly small. Notice that, by definition,

$$N_c \geq (c^{-1} q)^{(1+q+h)^{-1}}, \quad \text{with probability 1.} \qquad (4.2.7)$$

Based on the stopping rule in (4.2.6), we consider the *sequential point estimator* T_{N_c} whose risk is then equal to

$$\lambda_F^*(c) = E_F g(|T_{N_c} - \theta|) + c E_F N_c. \qquad (4.2.8)$$

T_{N_c} is termed an *asymptotic minimum risk estimator* (AMRE) of θ if

$$\lim_{c\downarrow 0}\{\lambda_F^*(c)/\lambda_F^0(c)\} = 1 \quad \forall F \in \mathcal{F}. \qquad (4.2.9)$$

Note that this asymptotic optimality of T_{N_c} is relative to its nonsequential counterpart based on the optimal n_c^0.

A general review of this AMRE, in a nonparametric setup, is given in Sen (1981d, §§ 10.5 and 10.6). Ghosh and Mukhopadhyay (1979) were the first to consider the case of $T_n = \bar{X}_n = n^{-1} \sum_{i=1}^{n} X_i$ for F not necessarily normal—their regularity conditions were relaxed by Chow and Yu (1981), while Sen and Ghosh (1981) treated the general case of U-statistics under full generality. Sen (1980b) treated earlier the case of R-estimators of location and Jurečková and Sen (1982) treated the case of M-estimators and L-estimators of location. Here, instead of reviewing these developments individually for these estimators, we intend to present the theory in a unified manner, and then to cite these cases as particular ones. Towards this goal, we make the following assumptions:

(I) With $g(\cdot)$ and q, defined as in (4.2.1) and (4.2.3),

$$g(|t_1 - t_2|) \leq K |t_1 - t_2|^{2q} \quad \forall \, t_1, t_2 \in R, \tag{4.2.10}$$

where K is a generic constant, independent of t_1, t_2.

(II) For some r (>4), there exist finite K_r and n_0, such that

$$E_F\{(n^{1/2} |T_n - \theta|)^r\} \leq K_r < \infty \quad \forall \, n \geq n_0. \tag{4.2.11}$$

(III) With q defined by (4.2.3), for every $n \geq n_0$,

$$n^q E\left\{\max_{m \,:\, |m-n| < \delta n} |T_m - T_n|^{2q}\right\} \to 0 \quad \text{as } \delta \downarrow 0. \tag{4.2.12}$$

(IV) There exists a τ (>0), such that for every $\varepsilon > 0$ and $n \geq n_0$,

$$P\{|d_n - \rho^*(F)| > \varepsilon\} \leq K(\varepsilon, \tau) n^{-1-\tau}, \quad K(\varepsilon, \tau) < \infty. \tag{4.2.13}$$

(V) Either (4.2.13) holds with $\tau > q + h$, or for every $n \geq n_0$,

$$P\left\{\sup_{m \geq n} |d_m - \rho^*(F)| > \varepsilon\right\} \leq K(\varepsilon, \tau) n^{-\tau} \quad \text{where } \tau > q + h, \tag{4.2.14}$$

and h (>0) is defined in (4.2.6).

For various nonparametric procedures, we shall comment on these conditions in the next section. Our basic theorems are the following.

THEOREM 4.2.1. *Under* (I) *through* (V), *the AMRE property in* (4.2.9) *holds.*

THEOREM 4.2.2. *Suppose that* $\{d_n\}$ *satisfy the following conditions:*

(i) *for some finite* β ($0 < \beta < \infty$), *depending on* F, *as* $n \to \infty$, *and*

$$n^{1/2}(d_n / \rho^*(F) - 1) \underset{\mathscr{D}}{\to} \mathscr{N}(0, \beta^2); \tag{4.2.15}$$

(ii) *for every* $\varepsilon > 0$,

$$\lim_{n \to \infty} P\left\{\max_{m \,:\, |m-n| < \delta n} n^{1/2} |d_m - d_n| > \varepsilon\right\} \downarrow 0 \quad \text{as } \delta \downarrow 0. \tag{4.2.16}$$

Then, whenever, in (4.2.6), h *is chosen to be greater than* $\frac{1}{2}$,

$$(n_c^0)^{-1/2}(N_c - n_c^0) \underset{\mathscr{D}}{\to} \mathscr{N}(0, (q+1)^{-2}\beta^2) \quad \text{as } c \downarrow 0. \tag{4.2.17}$$

In the rest of this section, we provide a brief outline of the proof of these theorems. Note that by (4.2.4), (4.2.6) and (4.2.13) (without requiring $\tau > q + h$),

$$N_c/n_c^0 \to 1, \quad \text{with probability 1} \quad \text{as } c \downarrow 0. \tag{4.2.18}$$

Further, by (4.2.6), $P(N_c > n) < P\{n^{q+1} < c^{-1}q(d_n + n^{-h})\}$, and hence, by (4.3.18) and (4.2.18), for every $\varepsilon > 0$, as $c \downarrow 0$,

$$(n_c^0)^{-1} E\{N_c I(N_c > n_c^0(1+\varepsilon))\} \leq (1+\varepsilon) P\{N_c \geq n_c^0(1+\varepsilon)\}$$

$$+ (n_c^0)^{-1} \sum_{n \geq n_c^0(1+\varepsilon)} P\{N_c > n\} \to 0 \tag{4.2.19}$$

(as for $n \geq n_c^0(1+\varepsilon)$, $n^{q+1}c/q \geq \rho^*(F)(1+\varepsilon)$). Since N_c is positive with probability 1, by (4.2.18) and (4.2.19), we conclude that

$$(E_F N_c)/n_c^0 \to 1 \quad \text{as } c \downarrow 0. \tag{4.2.20}$$

Hence, by virtue of (4.2.2), (4.2.8) and (4.2.20), to prove (4.2.9), it suffices to show that as $c \downarrow 0$,

$$|E_F g(|T_{N_c} - \theta|) - \rho_{n_c^0}(F)| = o(c^{q/(q+1)}). \tag{4.2.21}$$

On the other hand, by (4.2.10), (4.2.11), (4.2.18) and the Hölder inequality,

$$E_F\{g(|T_{N_c^0} - \theta|)I(|N_c - n_c^0| > \varepsilon n_c^0)\} = o(c^{q/(q+1)}) \quad \text{as } c \downarrow 0, \tag{4.2.22}$$

for every $\varepsilon > 0$, while by (4.2.10) and (4.2.12), as $c \downarrow 0$,

$$E_F\{|g(|T_{N_c} - \theta|) - g(|T_{n_c^0} - \theta|)|I(|N_c - n_c^0| \leq \varepsilon n_c^0)\} = o(c^{q/(q+1)}), \quad \varepsilon \downarrow 0. \tag{4.2.23}$$

Hence, to prove (4.2.21), it suffices to show that for every $\varepsilon > 0$, as $c \downarrow 0$,

$$E_F g(|T_{N_c} - \theta|)I(N_c < n_c^0(1-\varepsilon)) = o(c^{q/(q+1)}), \tag{4.2.24}$$

$$E_F g(|T_{N_c} - \theta|)I(N_c > n_c^0(1+\varepsilon)) = o(c^{q/(q+1)}). \tag{4.2.25}$$

Note that by virtue of (4.2.10), for every $m \geq n_c^0(1+\varepsilon)$,

$$E_F g(|T_{N_c} - \theta|)I(N_c \geq m) = E_F\left\{\sum_{n \geq m} I(N_c = n)g(|T_n - \theta|)\right\}$$

$$\leq K \sum_{n \geq m} E_F\{I(N_c = n)|T_n - \theta|^{2q}\}$$

$$\leq K\left(\sum_{n \geq m} E_F|T_n - \theta|^r\right)^{2q/r} (P\{N_c \geq m\})^{1-2q/r} \tag{4.2.26}$$

(by the Hölder type inequalities), where r is defined by (4.2.11). Since, by (4.2.13), $P\{N_c > n_c^0(1+\varepsilon)\} = O((n_c^0)^{-1-\tau})$, using (4.2.12) and (4.2.26), it is easy to show that (4.2.25) holds. In this context note that $0 < q \leq 1$, so that for $\tau \geq q$ (in (4.2.13)), one may relax (in (4.2.11)) r to $> 2 + q$. The treatment of (4.2.24) is very similar, where for $P\{N_c \leqq n_c^0(1-\varepsilon)\}$, we need to use (4.2.7) and (4.2.14). In this case also, r in (4.2.11) may be relaxed a bit, when, in (4.2.6), h (>0) is chosen small. These details are provided in Sen (1981d, §§ 10.5, 10.6), dealing with specific cases, and hence, are omitted.

Next, we may use (4.2.6) to have two opposite inequalities for N_c^{q+1} and $(N_c-1)^{q+1}$, so that for $h > \frac{1}{2}$, using (4.2.18), we have

$$((N_c/n_c^0)^{q+1}-1) = (d_{N_c}/\rho^*(F)-1) + o_p(c^{1/2(q+1)}) \quad \text{as } c \downarrow 0, \qquad (4.2.27)$$

while by (4.2.4), $(n_c^0)^{1/2} = O(c^{1/2(q+1)})$. Hence, by using (4.2.15) and (4.2.16), along with (4.2.18), we conclude that as $c \downarrow 0$,

$$(n_c^0)^{1/2}((N_c/n_c^0)^{q+1}-1) \underset{\mathcal{D}}{\to} \mathcal{N}(0; \beta^2), \qquad (4.2.28)$$

and (4.2.17) follows from (4.2.28) and the Slutzky theorem. Q.E.D.

4.3. Verification of regularity conditions: Nonparametric case. Let us first consider the case of U-statistics; we may refer to § 3.2 for the basic notions. We define the kernel (of degree $m \geqq 1$) as in (3.2.1), the U-statistic U_n as in (3.2.2). If in (4.2.1), $g(t)$ is taken as equal to t^2, then (4.2.3) holds with $q = 1$ and $\rho^*(F) = m^2\zeta_1$, where ζ_1 is defined by (3.2.10) (with $c = 1$). In this case, a suitable estimator of $\rho^*(F)$ is $m^2 s_n^2$, where s_n^2 is defined by (3.2.12). With these notations, we note that here (4.2.10) holds with $q = 1$, while (4.2.11) holds whenever the kernel has a finite absolute rth moment, for some $r > 4$. By virtue of the reversed martingale property of $\{U_n\}$, (4.2.12) holds whenever the kernel has a finite second moment. The estimator s_n^2 in (3.2.12) is expressible as a linear combination of several U-statistics (of degrees $m+1$ to $2m$), whose moments of order r^* exist whenever the kernel has a finite moment of order $2r^*$ ($r^* \geqq 1$). Thus, whenever r in (4.2.1) is >4, (4.2.13) holds with $\tau > 0$. See Sen and Ghosh (1981) for some of these details. On the other hand, by using the reversed martingale property of the individual U-statistics appearing in s_n^2, we may appeal to the Kolmogorov inequality and claim that (4.2.14) holds whenever the kernel has a finite rth order moment and $r/4 > \tau$. In fact, here, by the use of the reserved martingale property of $\{U_n\}$, one may even replace (4.2.11) by the finiteness of the moments of the kernel up to the order $2+\delta$, for some $\delta > 0$. For some of these details, we may refer to § 10.5 of Sen (1981d). If in (4.2.1), we let $g(t) = t$, then (4.2.3) holds with $q = \frac{1}{2}$ and $\rho^*(F) = \sqrt{2/\pi} m \zeta_1^{1/2}$, so that the results continue to hold, even under somewhat weaker moment conditions on the kernel. A similar treatment holds for $T_n = b(U_n)$, when $b(\cdot)$ satisfy some general smoothness conditions under which (4.2.10)–(4.2.14) hold.

Next, we consider the case of R-estimators of location, treated in Sen (1980). Here the X_i are i.i.d.r.v. with a d.f. F, given by

$$F(x) = F_0(x-\theta), \quad x \in E, \quad \theta \text{ unknown}, \quad F_0(x) + F_0(-x) \equiv 1. \quad (4.3.1)$$

For every n ($\geqq 1$) and real b, consider the *signed-rank statistic*

$$S_n(b) = \sum_{i=1}^{n} \text{sign}(X_i - b) a_n^*(R_{ni}^+(b)), \quad n \geqq 1, \quad b \in E, \qquad (4.3.2)$$

where $a_n^*(1) \le \cdots \le a_n^*(n)$ are suitable scores and $R_{ni}^+(b) = \text{rank}$ of $|X_i - b|$ among $|X_1 - b|, \ldots, |X_n - b|$, for $i = 1, \ldots, n$. $S_n(\theta)$ has a d.f. symmetric about 0 (and independent of F_0), while $S_n(b)$ is $\searrow$ in b. The R-estimator $\hat{\theta}_{n(R)}$ of θ, based on $X_1, \ldots, X_n$, is then defined by

$$\hat{\theta}_{n(R)} = \tfrac{1}{2}(\sup \{b : S_n(b) > 0\} + \inf \{b : S_n(b) < 0\}). \qquad (4.3.3)$$

$\hat{\theta}_{n(R)}$ is a median-unbiased, translation-invariant estimator of θ. If the scores are square integrable and F_0 possesses a density function f_0 with a finite Fisher information $I(f_0) = \int (f_0'/f_0)^2 \, dF_0 < \infty$, then

$$n^{1/2}(\hat{\theta}_{n(R)} - \theta) \sim \mathcal{N}(0, \nu^2) \qquad (4.3.4)$$

where ν^2 will be defined in the sequel. We define $a_n^*(k) = E\phi^+(U_{nk})$ or $\phi^+(k/(n+1))$, $k = 1, \ldots, n$ where $U_{n1} < \cdots < U_{nn}$ are the order r.v. of a sample of size n from the uniform $(0, 1)$ d.f., $\phi^+(u) = \phi((1+u)/2)$, $0 < u < 1$ and ϕ is a skew-symmetric and $\nearrow$ score function on $(0, 1)$. Then, we have

$$\nu^2 = A^2/\gamma^2, \quad A^2 = \int_0^1 \phi^2(u) \, du, \quad \gamma = \int_0^1 \phi(u)\{-f_0'(F_0^{-1}(u))/f_0(F_0^{-1}(u))\} \, du,$$

where ϕ is assumed to be square integrable. If we denote the ordered X_i by $X_{n,1} < \cdots < X_{n,n}$, then it follows from Sen (1980a) that there exists a sequence $\{k_n\}$ of positive integers, such that $X_{n,k_n} < \hat{\theta}_{n(R)} < X_{n,n-k_n+1}$, with probability 1, $\forall n \ge 1$, where as $n \to \infty$, $n^{-1}k_n \to \alpha : 0 < \alpha < \tfrac{1}{2}$. As such, if we assume that $E |X|^a < \infty$ for some $a < 0$ (a not necessarily ≥ 1), then using the moment results of sample quantiles (Sen (1959)), it follows that for every r (>0), there exists an n_r ($<\infty$), depending on a, ϕ and r, such that $E |\hat{\theta}_{n(R)}|^r < \infty$, $\forall n \ge n_r$, and moreover, $\lim_{n\to\infty} E |\hat{\theta}_{n(R)} - \theta|^r = 0$, $\forall r$ ($<\infty$). Let us now assume that for some generic constant K ($<\infty$),

$$\left| \frac{d^r}{du^r} \phi(u) \right| \le K\{u(1-u)\}^{-\delta-r}, \qquad 0 < u < 1, \quad r = 0, 1, 2, \qquad (4.3.5)$$

where $\delta < \tfrac{1}{6}$. (We may improve to $\delta < \tfrac{1}{4}$, provided the density f_0 satisfies some extra condition, see Sen (1980a).) Then, writing $\delta < (4+2\tau)^{-1}$, $\tau > 0$, we have for every $k < 2(1+\tau)$,

$$\lim_{n\to\infty} E\{n^{k/2} |\hat{\theta}_{n(R)} - \theta|^k\} = \nu^k E |Z|^k, \qquad (4.3.6)$$

where ν is defined as before and Z has a standard normal d.f. The proof of (4.3.6) exploits the linearity results on $S_n(b)$ to a greater extent and is given in Sen (1980b). It has been shown there that when θ holds,

$$n^{1/2}(\theta_{n(R)} - \theta)\gamma = n^{-1/2}S_n(\theta) + R_n, \qquad (4.3.7)$$

where $R_n \to 0$ a.s., as $n \to \infty$ and $E|R_n|^k \to 0$, $\forall k < 2(1+\tau)$. Moreover, it has been shown earlier by Sen and Ghosh (1971) that $\{S_n(\theta); n \ge 1\}$ is a martingale sequence, so that (4.2.11) and (4.2.12) follows from (4.3.6), (4.3.7) and Doob's submartingale inequality. Here, if in (4.2.1), we take $\rho(t) = t^2$, then $\rho^*(F) = \nu^2$

and $q = 1$, while for $\rho(t) \equiv t$, $\rho^*(F) = \sqrt{2/\pi}\nu$. Since A^2 is known, while γ is unknown, to estimate $\rho^*(F)$, it suffices to estimate γ. Since $S_n(\theta)$ has a known distribution, symmetric about 0 and $n^{-1/2}S_n(\theta) \sim \mathcal{N}(0, A^2)$, there exists a sequence $\{S_{n\alpha}\}$ of real (positive) values such that $P\{|S_n(\theta)| \leq S_{n\alpha}| \theta\} = 1 - \alpha_n$, $\alpha_n \to \alpha$ $(0 < \alpha < 1)$. Let then $\hat{\theta}_{n,L} = \sup\{b: S_n(b) > S_{n\alpha}\}$, $\hat{\theta}_{n,U} = \inf\{b: S_n(b) < -S_{n\alpha}\}$ and let

$$d_n = 2S_{n\alpha}\{n(\hat{\theta}_{n,U} - \hat{\theta}_{n,L})\}^{-1}, \qquad n \geq 3. \tag{4.3.8}$$

Then, using again the results of Sen (1980a), it follows that with δ defined as in (4.3.5), for every $\varepsilon > 0$, there exists a $c(\varepsilon) < \infty$, such that

$$P\{|d_n/\gamma - 1| > \varepsilon\} \leq c(\varepsilon)n^{-s} \quad \forall n \geq n_\varepsilon \tag{4.3.9}$$

where $s = 1 + \tau$, $\delta < (4 + 2\tau)^{-1}$, $\tau > 0$. Thus, (4.2.13) and (4.2.14) hold. This shows that Theorem 4.2.1 holds for R-estimators under the above mentioned regularity conditions, and these conditions hold for a broad class of d.f.'s and for almost all the commonly used scores. Verification of (4.2.15) and (4.2.16) demands stronger form of linearity results on the $S_n(b)$. Some of these have been studied, under additional regularity conditions, by Hušková and Jurečková (1981) and Hušková (1982), among others, and these may be incorporated in our study too.

Let us next consider the case of M-estimators of location. Consider a *score function* $\psi = \{\psi(x): x \in E\}$, such that

$$\psi(x) = \psi_1(x) + \psi_2(x), \qquad x \in E, \tag{4.3.10}$$

where ψ_1 and ψ_2 are both $\nearrow$ and skew-symmetric functions; ψ_1 is absolutely continuous on any bounded interval in E and ψ_2 is a step function having finitely many jumps. In addition, we assume that $\psi(x) = \psi(c)\operatorname{sign} x$ for $|x| \geq c$ (>0), where c $(<\infty)$ is a given constant, and further that ψ is nonconstant, so that

$$0 < \sigma_\psi^2 = \int_{-\infty}^{\infty} \psi^2(x)\, dF_0(x) < \infty. \tag{4.3.11}$$

On the d.f. F_0, we have essentially the same assumptions as in the case of R-estimators, and we define

$$\gamma_M = \int_{-\infty}^{\infty} \psi(x)\{-f_0'(x)/f_0(x)\}\, dF_0(x). \tag{4.3.12}$$

Note that unlike the case of R-estimators, here both σ_ψ^2 and γ_M are unknown (functionals of the unspecified d.f. F_0). Let then

$$M_n(t) = \sum_{i=1}^{n} \psi(X_i - t), \qquad t \in E, \tag{4.3.13}$$

which is a $\searrow$ function of $t \in E$. Also, $M_n(\theta)$ has a d.f. symmetric about 0. Hence, we may define an M-estimator of θ by

$$\theta_{n(M)} = \tfrac{1}{2}(\sup\{t: M_n(t) > 0\} + \inf\{t: M_n(t) < 0\}). \tag{4.3.14}$$

Then, for the given class of ψ, $\hat{\theta}_{n(M)}$ is a median-unbiased, translation-invariant estimator of θ, and as $n \to \infty$,

$$n^{1/2}(\hat{\theta}_{n(M)} - \theta) \sim \mathcal{N}(0, \nu^2), \qquad \nu^2 = \sigma_\psi^2 / \gamma_M^2. \qquad (4.3.15)$$

Since for every (fixed) t, $M_n(t)$ involves a sum of independent and bounded valued r.v.'s, using a Hoeffding (1963) type inequality, Jurečková and Sen (1982) have verified (4.3.6) for $\hat{\theta}_{n(M)}$. Also, an estimator of σ_ψ^2, they considered

$$s_n^2 = \int_{-\infty}^{\infty} \psi^2(x - \hat{\theta}_{n(M)}) \, d\hat{F}_n(x) = \frac{1}{n} \sum_{i=1}^{n} \psi^2(X_i - \hat{\theta}_{n(M)}). \qquad (4.3.16)$$

As before in (4.3.8), we equate $M_n(t)$ to $\pm\sqrt{n}\tau_{\alpha/2}s_n$ (where τ_ε is the upper $100\varepsilon\%$ point of the standard normal d.f.) and the solutions (for t) are denoted by $\hat{\theta}_{n,U}^M$ and $\hat{\theta}_{n,L}^M$, respectively. Then, parallel to (4.3.8), here we have the estimator

$$d_n = (2\tau_{\alpha/2}s_n)/\{\sqrt{n}(\hat{\theta}_{n,U} - \hat{\theta}_{n,L}^M)\}. \qquad (4.3.17)$$

If we define, side by side, $s_n^{02} = n^{-1} \sum_{i=1}^{n} \psi^2(X_i - \theta)$, by virtue of the boundedness of ψ and the Hoeffding (1963) inequality, for every $\varepsilon > 0$, there exists a $\rho(\varepsilon) : 0 < \rho(\varepsilon) < 1$, such that

$$P\{|s_n^{02} - \sigma_\psi^2| > \varepsilon\} \leq 2[\rho(\varepsilon)]^n \quad \forall n \geq n_0. \qquad (4.3.18)$$

Further, by (4.3.10) and (4.2.16),

$$|s_n^2 - s_n^{02}| \leq 2 \left\{ \frac{1}{n} \sum_{i=1}^{n} |\psi_1^2(X_i - \hat{\theta}_{n(M)}) - \psi_1^2(X_i - \theta)| \right.$$
$$\left. + \frac{1}{n} \sum_{i=1}^{n} |\psi_2^2(X_i - \hat{\theta}_{n(M)}) - \psi_2^2(X_i - \theta)| \right\}. \qquad (4.3.19)$$

By the absolute continuity of ψ_1 (or ψ_1^2) and the exponential rate of convergence of $\hat{\theta}_{n(M)}$ (to θ), the first term on the right-hand side of (4.3.19) converges to 0, with a probability converging at an exponential rate, while for the second term, the Bahadur (1966) representation of sample quantiles and the finiteness of the number and magnitudes of the jumps of ψ_2, a similar exponential rate is available. Thus, (4.3.18) holds with s_n^{02} being replaced by s_n^2 and the factor 2 on the right-hand side being replaced by some $c\ (<\infty)$. For the convergence of d_n to γ_M, one may appeal to the asymptotic linearity results in Jurečková and Sen (1981a, b) and conclude that here (4.3.9) holds with s arbitrarily chosen. These also yield, parallel to (4.3.7) that when θ holds,

$$n^{1/2}(\hat{\theta}_{n(M)} - \theta)\gamma_M = n^{-1/2}M_n(\theta) + R_n, \qquad (4.3.20)$$

where $E|R_n|^k \to 0$, $\forall k\ (>0)$, as $n \to \infty$ (and $R_n \to 0$ a.s., as $n \to \infty$). Since $\{M_n(\theta); n \geq 1\}$ is a martingale, as in the case of R-estimators, here also, (4.2.11)–(4.2.14) can be verified on the same line. Further, by virtue of the assumed boundedness of ψ, here for s_n^2 (through s_n^{02} and (4.3.19)) and d_n

(through (4.3.17)), (4.2.15)–(4.2.16) can be verified—we may refer to Jurečková and Sen (1981b) for these details. Thus, here both Theorems 4.2.1 and 4.2.2 hold. In passing, we may remark that as in the case of R-estimators, here also it suffices to assume that $E|X|^a < \infty$, for some $a > 0$. If, however, we are able to choose a larger than 2 (or 4), then the boundedness condition on ψ may be replaced by a suitable moment condition: $\int |\psi|^r \, dF < \infty$ for some $r > 2$ (or 4) under which Theorem 4.2.1 (or 4.2.2) holds, without requiring ψ to be bounded. However, in robust estimation, the boundedness of ψ is generally imposed to curb the influence of outliers, and hence, the higher order moment condition on F_0 is not that really needed.

L-estimators of location are also of considerable interest. If $X_{n,1} < \cdots < X_{n,n}$ stand for the ordered r.v.'s, corresponding to $X_1, \ldots, X_n$ (ties neglected with probability 1, by virtue of the assumed continuity of F), then, typically, an L-estimator of location is of the form

$$\hat{\theta}_{n(L)} = \sum_{i=1}^{n} c_{n,i} X_{n,i} \quad c_{n,i} \geq 0, \quad \forall i, \quad \sum_{i=1}^{n} c_{n,i} = 1, \tag{4.3.21}$$

where the scores $c_{n,i}$ are usually taken as $\phi_n(i/(n+1))$, $1 \leq i \leq n$, $\phi_n(u)$, $u \in (0, 1)$ being a suitable score function. For symmetric F_0, one additionally takes $c_{n,i} = c_{n,n-i+1}$, $1 \leq i \leq n$. Jurečková and Sen (1982) considered the class of all such $\hat{\theta}_{n(L)}$ for which for some $\alpha_0(0 < \alpha_0 < \frac{1}{2})$, $c_{n,i} = c_{n,n-i+1} = 0$, $\forall i \leq k_n$, where $n^{-1}k_n \to \alpha_0$ as $n \to \infty$. In this case, (4.3.6) holds by the usual moment convergence results on central order statistics, where ν^2 is defined by

$$\nu^2 = \iint_{E^2} \{F_0(x \wedge y) - F_0(x)F_0(y)\} \phi(F_0(x))\phi(F_0(y)) \, dF_0(x) \, dF_0(y),$$

$$\tag{4.3.22}$$

where $\phi(u) = \lim_{n \to \infty} \phi_n(u)$, $0 < u < 1$. A natural estimator d_n of ν^2, studied by Sen (1978), may be obtained by simply replacing $F_0(x)$ and $F_0(y)$ in (4.2.22) by the empirical d.f. $\hat{F}_n(x) = n^{-1} \sum_{i=1}^{n} I(X_i - \hat{\theta}_{n(L)} \leq x)$, $x \in E$, and its almost sure convergence properties were studied in Sen (1978) by using some reversed martingale characterizations. This leads to an easy verification of (4.2.13)–(4.2.14) where again one may choose τ (> 0) arbitrarily (as ϕ is bounded). To establish (4.2.12), Jurečková and Sen (1982) considered a representation, similar to (4.3.20), where $M_n(\theta)$ and γ_M are replaced by a sum of i.i.d.r.v.'s and a positive scale factor, respectively. Thus, Theorem 4.2.1 holds. Verification of (4.2.15) and (4.2.16) can be made using the results of Gardiner and Sen (1979) and Sen (1984c). Again, as $\phi(u) = 0$ in the two tails, for Theorems 4.2.1 and 4.2.2 to hold, we just need that $E|X|^a < \infty$, for some $a > 0$. For $a \geq 2$ and for some further condition on ϕ, one may eliminate the condition that $c_{n,i} = c_{n,n-i+1} = 0$, $\forall i \leq k_n$. However, from robustness considerations, this does not seem to be a very major issue.

4.4. Some generalized sequential point estimation procedures. Consider a two-sample problem: Let $\{X_i, i \geq 1\}$ be i.i.d.r.v. with a d.f. F, defined on E, and let $\{Y_j, j \geqq 1\}$ be a second (independent sequence) of i.i.d.r.v. with a d.f. G, also defined on E. In the classical two-sample location model, one assumes that

$$G(x) = F(x - \Delta), \qquad x \in E, \quad \Delta \text{ real.} \tag{4.4.1}$$

and one may proceed to estimate the (shift-) parameter Δ with a minimum risk, in the same sense as in § 4.2. Let $T_n = T(X_1, \ldots, X_n; Y_1, \ldots, Y_n)$ be an estimator of Δ based on samples of sizes n from each of the two populations, and assume that there exists an $n_0 (\geqq 1)$, such that for every $n \geqq n_0$, $\nu_n^2 = 2nE(T_n - \Delta)^2$ exists and $\nu_n^2 \to \nu^2 : 0 < \nu < \alpha$, as $n \to \infty$. (Instead of the mean square, we may take, as in § 4.2, some other metric too.) As in (4.2.2), we have the risk $\lambda_n(c; F, G) = 2cn + (2n)^{-1}\nu_n^2$, $n \geqq n_0$, and our objective is to minimize this risk by a proper choice of n. Since $\{\nu_n^2\}$ (as well as ν^2) depend on the unknown F, this minimum risk estimator is, in general, dependent on F (and hence, no fixed sample size procedure may achieve this goal, for all F belonging to a class $\mathcal{F}$.) As in § 4.2, we assume that there exists a sequence $\{\hat{\nu}_n^2\}$ of consistent estimates of ν^2, so that parallel to (4.2.6), we may define here a *stopping number* N_c by letting

$$N_c = \min \{n \geqq n_0 : 2n \geqq c^{-1/2}(\hat{\nu}_n + n^{-h})\}. \tag{4.4.2}$$

(Note that here $q = 1$.) We consider then the sequential estimator T_{N_c}, and denote its risk by $\lambda_F^*(c) = 2cEN_c + E(T_{N_c} - \Delta)^2$. Our goal is to establish the AMRE property in (4.2.9) in the present context too.

Sen (1983c) has considered, for T_n, a general class of two-sample R-estimators, and, under regularity conditions, parallel to the one-sample case (see § 4.3), has shown that Theorem 4.2.1 holds for such R-estimators as well. Earlier, Ghosh and Mukhopadhyay (1979), under comparatively stringent moment conditions, have studied similar results for the difference of the two sample means. We shall comment more on it in the sequel.

In the two-sample problem, one may have a somewhat more general setup, where

$$F(x) = F_0(x - \theta_1), \quad F(x) = G_0(x - \theta_2), \quad x \in E, \quad \Delta = \theta_1 - \theta_2, \tag{4.4.3}$$

where F_0 and G_0 are both symmetric d.f.'s, but they need not be of the same form. Here Δ is the difference of the two location parameters θ_2 and θ_1. In the so-called *Behrens–Fisher model*, one takes $G_0(x) = F_0(\nu x)$, for some (unknown) $\nu : 0 < \nu < \infty$. Under (4.4.3) or even the Behrens–Fisher model, the equal sampling scheme (from both F and G) may not lead to the MRE. Asymptotic properties of the stopping time N_c in (4.4.2) and the allied estimator T_{N_c}, under the general model in (4.4.3), have been studied in Sen (1983c) for a general class of R-estimators. In general, under (4.4.3), T_{N_c} may not have the AMRE property. To overcome this drawback, alternative sequential R-estimators (based on the difference of the one-sample R-estimators) were considered, where the two sample sizes are chosen in such a way that the AMRE property may hold.

Suppose that in (4.3.4), we denote ν^2 by ν_F^2 and ν_G^2, respectively, for the X

and Y sample observations. Also, defining γ as in after (4.3.4), we denote this by γ_F and γ_G, for the X and Y observations, respectively. Then, if one uses the same score function (ϕ) in both the cases, we have

$$v_F^2 = A^2/\gamma_F^2, \qquad v_G^2 = A^2/\gamma_G^2, \qquad (4.4.4)$$

and as $c \downarrow 0$, the optimal (nonsequential) sample sizes and the asymptotic minimum risk are given by

$$n_{1c}^2 \gamma_F^2 \sim n_{2c}^2 \gamma_G^2 \sim c^{-1}, \qquad \rho_{n_{1c}n_{2c}} \sim 2c^{1/2}(\gamma_F + \gamma_G). \qquad (4.4.5)$$

As such, if, as in (4.3.8), we define d_n^X and d_n^Y for the two samples, and let $V_n^X = A/d_n^X$ and $V_n^Y = A/d_n^Y$, then, we may define a pair of stopping numbers (N_{1c}, N_{2c}) by letting

$$N_{1c} = \min\{n \geq n_0 : n \geq c^{-1/2}(V_n^X + n^{-h})\},$$
$$N_{2c} = \min\{n \geq n_0 : n \geq c^{-1/2}(V_n^Y + n^{-h})\}, \qquad (4.4.6)$$

where, as in (4.2.6), $h\ (>0)$ is a suitable constant. The sequential estimator is then $\hat{\theta}_{N_{1c}(R)}^X - \hat{\theta}_{N_{2c}(R)}^Y$, where $\hat{\theta}_n^X$ and $\hat{\theta}_n^Y$ are defined as in (4.3.3) for the X and Y samples, respectively. The risk of this sequential estimator is the sum of the risks for the two individual sample sequential estimators, and hence, the results in § 4.3 may be incorporated to derive the desired AMRE property. For details, we may refer to Sen (1983c).

A somewhat more complicated situation arises with the *generalized U-statistics*. For an estimable parameter $\theta(F, G)$ (a functional of (F, G)), in a nonsequential setup, the optimal estimator is the corresponding generalized U-statistic $U_{n_1 n_2}$. For simplicity, we consider a *kernel* $\phi(X_i, Y_j)$ of degree $(1, 1)$ (so that $\theta(F, G) = E\phi(X_i, Y_j) = \iint \phi(x, y)\, dF(x)\, dG(y))$, so that $U_{n_1 n_2} = (n_1 n_2)^{-1} \sum_{i=1}^{n_1} \sum_{j=1}^{n_2} \phi(X_i, Y_j)$, $n_1 \geq 1$, $n_2 \geq 1$. (As an example, consider the Wilcoxon–Mann–Whitney statistic, for which $\phi(X, Y)$ is 1 or 0 according as X is $\leq$ or $> Y$.) Note that $EU_{n_1 n_2} = \theta(F, G)$ and

$$E(U_{n_1 n_2} - \theta(F, G))^2 = n_1^2 \zeta_{10} + n_2^2 \zeta_{01} + (n_1 n_2)^{-1} \zeta_{11}, \qquad (4.4.7)$$

where ζ_{10}, ζ_{01} and ζ_{11} are (unknown) parameters (these are also regular functionals of F and G). As such, if we consider the risk $\rho_c(n_1, n_2) = E(U_{n_1 n_2} - \theta(F, G))^2 + c(n_1 + n_2)$, that will depend on $\zeta_{10}, \zeta_{01}, \zeta_{11}, n_1, n_2$ and $c\ (>0)$, and no fixed sample size solution may lead to the MRE, simultaneously for all F, G belonging to a suitable class. Ghosh and Sen (1984) have considered AMRE (sequential versions) of $U_{n_1 n_2}$ based on jackknifed estimators of the parameters ζ_{10} and ζ_{01}. Unlike the case of R-estimators, here these jackknifed estimators depend on both the sample observations, and hence, some special care needs to be taken in defining the stopping rule in a precise manner. Directional reversed martingale property of $U_{n_1 n_2}$ (viz. Sen (1974a)) and some maximal inequalities have been used in this context to establish the desired uniform integrability results; these are the natural generalizations of the one-sample results, treated earlier by Sen and Ghosh (1981). Asymptotic

normality of the stopping time for this generalized model has also been studied
by Ghosh and Sen (1984).

Let us now consider the multivariate location model, treated in Sen (1984f).
Let $\{\mathbf{X}_i; i \geqq 1\}$ be a sequence of i.i.d.r.v.'s with a d.f. $\mathcal{F}$, defined on E^p, for some
$p \geqq 1$. We assume that

$$F(\mathbf{x}) = F_0(\mathbf{x} - \boldsymbol{\theta}), \quad \mathbf{x} \in E^p, \quad F_0 \text{ diagonally symmetric}, \quad (4.4.8)$$

where $\boldsymbol{\theta} = (\theta_1, \ldots, \theta_p)'$ is the location parameter (vector) of $\mathcal{F}$. Let $\mathbf{T}_n = (T_{n1}, \ldots, T_{np})'$ be a suitable estimator of $\boldsymbol{\theta}$ (based on $\mathbf{X}_1, \ldots, \mathbf{X}_n$) and consider
a *risk* function

$$\lambda_F(n, c) = E\rho(\mathbf{T}_n, \boldsymbol{\theta}) + cn, \quad c > 0, \quad (4.4.9)$$

where $\rho(\mathbf{a}, \mathbf{b})$ is a suitable metric defined on $E^p \times E^p$. Typically, one may take
$\rho(\mathbf{a}, \mathbf{b}) = \max_{1 \leqq j \leqq p} |a_j - b_j| = \|\mathbf{a} - \mathbf{b}\|$, or $p^{-1} \sum_{j=1}^{p} |a_j - b_j|$ or $(\mathbf{a} - \mathbf{b})'\mathbf{Q}(\mathbf{a} - \mathbf{b})$ with
some p.d. $\mathbf{Q}$. Again, the basic problem is to choose an optimal n, leading to the
minimum value of the risk in (4.4.9); this optimal n, in general, depends on the
unknown F_0, through $E\rho(\mathbf{T}_n, \boldsymbol{\theta})$, and hence, sequential procedures are solicited
to obtain an optimal solution (in some convenient setup, where $c \to 0$). As in
(4.2.3), we assume here that for some $q: 0 < q \leqq 1$, $n^q E\rho(\mathbf{T}_n, \boldsymbol{\theta}) \to \delta(F_0)$, as
$n \to \infty$, and, further, there exists a sequence $\{d_n\}$ of consistent estimators of
$\delta(F_0)$. Then, one may define a *stopping variable*

$$N_c = \min\{n \geqq n_0 : n^{1+q} \geqq c^{-1}q(d_n + n^{-h})\}, \quad c > 0, \quad (4.4.10)$$

where $h\ (>0)$ is a suitable constant. Then $\mathbf{T}_{N_c}$ is the desired sequential
estimator of $\boldsymbol{\theta}$, and one is interested in the (asymptotic) properties of $\mathbf{T}_{N_c}$,
including its AMRE property (when $c \downarrow 0$). In this context, one would make
assumptions very similar to (4.2.11) through (4.2.14); in addition, we assume
that $\delta(F_0) = \delta(\boldsymbol{\Gamma})$, where $\boldsymbol{\Gamma} = \lim_{n \to \infty} \{n E_F(\mathbf{T}_n - \boldsymbol{\theta})(\mathbf{T}_n - \boldsymbol{\theta})'\}$. Then, some natural
estimates $\hat{\boldsymbol{\Gamma}}_n$ of $\boldsymbol{\Gamma}$ may be incorporated to define d_n (viz. $d_n = \delta(\hat{\boldsymbol{\Gamma}}_n)$, $n \geqq n_0$). If
$n^{1/2}(\hat{\boldsymbol{\Gamma}}_n - \boldsymbol{\Gamma})$ has asymptotically a multinormal distribution and $\{\hat{\boldsymbol{\Gamma}}_n\}$ satisfy the
uniform continuity (in probability), with respect to $n^{1/2}$, then under some
smoothness conditions on $\delta(\cdot)$, such that the Lipschitz-type condition, (4.2.15)–
(4.2.16) can be established as well. Thus, Theorems 4.2.1 and 4.2.2 extend
naturally to the multivariate case.

For the multivariate R-estimators of location, under the same regularity
conditions as in § 4.3, the regularity conditions have all been shown to hold by
Sen (1984f). The basic approach is a natural extension of the one in (4.3.3)
through (4.3.9) (to the multivariate case); however, unlike the univariate case,
here, the vector of (coordinatewise) signed rank statistics (in (4.3.2)) has an
unknown dispersion matrix, which we need to estimate to define $\hat{\boldsymbol{\Gamma}}_n$. Strong
consistency results and rates of convergence for this estimator are studied in
Sen (1984f). It has also been indicated there that parallel results for the
multivariate M- and L-estimators hold under the same regularity conditions as
in § 4.3.

For the estimation of the mean vector of a multinormal distribution,

James–Stein (1961) type (and related shrinkage) estimators are known to have smaller risks than the classical maximum likelihood estimator (in the non-sequential case). Recently, Ghosh and Sen (1983) have considered a two-stage James–Stein type estimator for the same problem; sequential analogues are being investigated now. It would be of natural interest to consider nonparametric counterparts of such sequential procedures. Sen (1984d) and Sen and Saleh (1984), (1985) have considered some (nonsequential) shrinkage R-estimators in the multivariate case, and their sequential versions are under investigation.

4.5. Sequential confidence regions: Nonparametric case. As in § 4.2, consider a sequence $\{X_i; i \geq 1\}$ of i.i.d.r.v. with a d.f. $\mathcal{F}$, and let $\theta\, (= \theta(F))$ be a parameter of interest. Based on a sample of size n, it is desired to locate two statistics L_n and U_n, such that (i) $L_n \leq U_n$ a.e., and (ii) when θ holds, $L_n \leq \theta \leq U_n$ hold with a probability greater than or equal to $1-\alpha$ $(0 < \alpha < 1)$; here, (L_n, U_n) is the *confidence interval* for θ, $1-\alpha$ is the *confidence coefficient* (coverage probability), L_n and U_n are the *lower and upper confidence limits* and $d_n = U_n - L_n$ is the *width* of the confidence interval. In practical problems, we often need to determine such a confidence interval with the additional property that for given $1-\alpha$, n is so chosen that

$$0 \leq d_n = U_n - L_n \leq 2d \quad \text{for some given } d > 0. \tag{4.5.1}$$

For normal F with location θ and unknown variance σ^2, the nonexistence of a fixed sample size procedure for this problem was proved by Dantzig (1940), while Stein (1945) considered a two-stage (sequential) procedure which satisfies the above requirements, though it may not be fully efficient. For a general class of d.f.'s admitting finite second moments, Chow and Robbins (1965) considered a modification of the Stein procedure, where the sample variance is updated in a sequential scheme to yield the full asymptotic efficiency (when $d \downarrow 0$). The asymptotic nonparametric character of the Chow–Robbins approach has been fully exploited in various directions, where θ need not be the location parameter and (L_n, U_n) need not be based on the sample means and variances. A general account of these developments is given in Sen (1981d, Chap. 10). In the nonparametric case, we may, essentially, take two different types of approaches to this problem. First, as in the normal theory case, one may consider suitable estimators $\{T_n\}$ of θ along with their estimated variances $\{s_n^2\}$, and base a procedure on these statistics—this we term a Type A procedure. Secondly, one may consider suitable (rank or M-) statistics and obtain by alignment confidence intervals from them as well—this is termed a Type B procedure. For either procedure, we have a well defined *stopping rule* on which the sequential procedure rests and satisfies the desired properties.

Consider the Type A procedure first. Let $\{T_n\}$ be a sequence of (point) estimators of θ, such that there exist a (known) nondecreasing function $\psi(n)$ (of n on $\{1, 2, \ldots, \infty\}$) and a (possibly unknown) parameter σ_θ^2 $(0 < \sigma_\theta < \infty)$, such that

$$\psi(n)[T_n - \theta]/\sigma_\theta \sim \mathcal{N}(0, 1) \quad \text{as } n \to \infty. \tag{4.5.2}$$

Also, assume that there exists a sequence $\{s_n^2\}$ of consistent estimators of σ_θ^2. Let then n_0 (≥ 2) be an initial sample size, and define a stopping variable $N(d)$ by letting

$$N(d) = \min\{n \geq n_0 : s_n \leq \psi(n)d/\tau_{\alpha/2}\}, \qquad d > 0, \tag{4.5.3}$$

where τ_ε is the upper $100\varepsilon\%$ point of the standard normal d.f. Then, the desired (sequential) confidence interval for θ is

$$I_{N(d)} = [T_{N(d)} - d, \, T_{N(d)} + d], \qquad d > 0. \tag{4.5.4}$$

Note that $N(d)$ is well defined for every $d > 0$. If we assume that

$$s_n^2 \to \sigma_\theta^2 \quad \text{a.s. as } n \to \infty, \tag{4.5.5}$$

then $N(d)$ is finite a.s. for every $d > 0$ and $N(d)$ is $\searrow$ in d (>0). The procedure is termed *asymptotically (as $d \downarrow 0$) consistent* if

$$\lim_{d \downarrow 0} P\{\theta \in I_{N(d)}\} = 1 - \alpha. \tag{4.5.6}$$

Towards this, we define (for every $d > 0$), $n(d) = \min\{n \geq n_0 : \sigma_\theta \leq d\psi(n)/\tau_{\alpha/2}\}$. Note that, typically, $\psi(n) = \sqrt{n}$. In general, we assume that for every $a \in (0, \infty)$,

$$\lim_{n \to \infty} \psi(an)/\psi(n) = s(a) \quad \text{is } \uparrow \text{ in } a : s(1) = 1. \tag{4.5.7}$$

Then, note that by (4.5.3), $s_{N(d)-1} > \psi(N(d)-1)d/\tau_{\alpha/1} \geq [\psi(N(d)-1)/\psi(N(d))]s_{N(d)}$, so that by (4.5.5) and (4.5.7), we conclude that

$$N(d)/n(d) \to 1 \quad \text{a.s., as } d \downarrow 0. \tag{4.5.8}$$

Hence, by virtue of (4.5.2), (4.5.8) and the definition of $n(d)$, to prove (4.5.6), it suffices to verify the usual (Anscombe) "uniform continuity in probability" condition: For every $\varepsilon > 0$ and $\eta > 0$, there exist an n^0 and a δ (>0), such that

$$P\left\{\max_{m:|m-n|\leq\delta n} |T_m - T_n| > \varepsilon/\psi(n)\right\} < \eta \quad \forall\, n \geq n^0. \tag{4.5.9}$$

Now, whenever $\{\psi(n)[T_m - \theta]/\sigma_\theta; m \geq n\}$ converges weakly to a Gaussian process (on $[0, 1]$) or $\{\psi^2(k)[T_k - \theta]/\psi(n); k \leq n\}$ to a Gaussian process on $[0, 1]$, the tightness part of this invariance principle ensures (4.5.9). For various nonparametric statistics, the invariance principles studied in Sen (1981d, Part I) thus enable us to verify (4.5.9) under minimal regularity conditions. Again, note that by (4.5.3), for every $d > 0$,

$$\psi(N(d)-1) \leq \psi(n_0-1)I_{[N(d)=n_0]} + d^{-1}\tau_{\alpha/2}s_{N(d)}I_{[N(d)>n_0]}, \tag{4.5.10}$$

where

$$s_{N(d)}I_{[N(d)n_0]} = \sum_{n>n_0} s_n I_{[N(d)=n]} \leq \sup_{n\geq n_0} s_n. \tag{4.5.11}$$

Hence, if either $E\{\sup_{n\geq n_0} \psi^{-1}(s_n)\} < \infty$ or $\sum_{n>n_0} E\{\psi^{-1}(s_n)I_{[N(d)=n]}\} < \infty$, we will have $EN(d) < \infty$; using this along with (4.5.8), one obtains that

$$\lim_{d \downarrow 0} EN(d)/n(d) = 1; \tag{4.5.12}$$

this is termed the *asymptotic (as $d \downarrow 0$) efficiency* of the procedure. In the case of $T_n = \bar{X}_n$, $n \geq 1$, $s_n^2 = (n-1)^{-1} \sum_{i=1}^{n} (X_i - \bar{X}_n)^2$, $n \geq 2$ is a reverse martingale, so that $E\{\sup_{n \geq n_0} s_n^2\} < \infty$ whenever $E_F\{s_{n_0}^2 \cdot (\log s_{n_0}^2)\} < \infty$. In this case, one may even take the minimal assumption of finite second moment by using some other inequalities (see Sen (1981d, § 10.2)); for general U-statistics, s_n^2, the jackknifed estimator, can be expressed as a linear combination of several reversed martingales, so that if for some $r > 2$, the rth moment of the kernel exists, then $E(\sup_{n \geq n_0} s_n^2) < \infty$. The weak invariance principles for such U-statistics (again based on the reversed martingale characterizations) have been studied in Sen (1981d, Chap. 3), so that (4.5.6) and (4.5.12) hold for the entire class of U-statistics (and related von Mises' functionals) under the conditions that $\theta(F)$ is stationary of order 0 and the kernel has finite (absolute) moment or order r, for some $r > 2$.

The confidence interval in (4.5.4) can be extended to the multiparameter case, where $\boldsymbol{\theta}$ and $\{\mathbf{T}_n\}$ are appropriate p-vectors, for some $p \geq 1$. in that case, in place of (4.5.2), we need the asymptotic multinormality of $\psi(n)(\mathbf{T}_n - \boldsymbol{\theta})$ (with mean $\mathbf{0}$ and dispersion matrix $\boldsymbol{\Sigma}$), while in (4.5.5), we need to have $\{\mathbf{S}_n\}$, such that $\mathbf{S}_n \to \boldsymbol{\Sigma}$ a.s., as $n \to \infty$. In (4.5.3), we take

$$N(d) = \min \{n \geq n_0 : \mathrm{ch}_1(\mathbf{S}_n) \leq d^2 \psi^2(n)/\chi_{p,\alpha}^2\}, \qquad d > 0, \qquad (4.5.13)$$

where $\chi_{p,\alpha}^2$ is the upper $100\alpha\%$ point of the chi-square d.f. with p degrees of freedom and $\mathrm{ch}_1(\mathbf{A}) = $ largest characteristic root of $\mathbf{A}$. Then, replacing (4.5.4) by

$$I_{N(d)} = \{\boldsymbol{\theta} : |\mathbf{l}'(\mathbf{T}_n - \boldsymbol{\theta})| \leq (\mathbf{l}'\mathbf{l})^{1/2} d, \forall \mathbf{l} \neq \mathbf{0}\}, \qquad (4.5.14)$$

we conclude that under the same regularity conditions on $\{\mathbf{T}_n\}$ and $\{\mathbf{S}_n\}$, (4.5.6) and (4.5.12) hold, with

$$n(d) = \min \{n \geq n_0 : \mathrm{ch}_1(\boldsymbol{\Sigma}) \leq d^2 \psi^2(n)/\chi_{p,\alpha}^2\}, \qquad d > 0. \qquad (4.5.15)$$

Let us next consider Type B sequential confidence regions. Again, for simplicity, we consider first the case of a single θ. Consider the one-sample location model and, as in before (4.3.8), define $\hat{\theta}_{n,L}$ and $\hat{\theta}_{U,n}$; then $\{\hat{\theta}_{L,n} \leq \theta \leq \hat{\theta}_{U,n}\}$ is a distribution-free confidence interval for θ with confidence coefficient $1 - \alpha_n$. In this scheme, we choose α_n such that $1 - \alpha_n$ is the smallest admissible value $\geq 1 - \alpha$. Let then

$$N(d) = \min \{n \geq n_0 : \hat{\theta}_{U,n} - \hat{\theta}_{L,n} \leq 2d\}, \qquad d > 0. \qquad (4.5.16)$$

Note that by definition of $N(d)$, $I_{N(d)} = [\hat{\theta}_{L,n}, \hat{\theta}_{U,n}]$ has width $\leq 2d$, so we need to verify (4.5.6) (when $\alpha_n \to \alpha$ as $n \to \infty$). In this context, the asymptotic linearity results on $S_n(b)$ (in b around θ) play a fundamental role. Sen and Ghosh (1971) considered the a.s. behavior of $n^{1/2}(\hat{\theta}_{U,n} - \hat{\theta}_{L,n})\gamma$ and verified that as $n \to \infty$,

$$\gamma n^{1/2}(\hat{\theta}_{U,n} - \hat{\theta}_{L,n}) \to 2A\tau_{\alpha/2} \quad \text{a.s.}, \qquad (4.5.17)$$

where γ and A are defined as in after (4.3.4). (Note that A is known, but γ is an unknown parameter.) If γ were known, by virtue of (4.3.4) and (4.5.17), if we had taken

$$n(d) = [A^2 \tau_{\alpha/2}^2 / \gamma^2 d^2] + 1, \qquad d > 0, \tag{4.5.18}$$

then for $n = n(d)$, (4.5.6) would be true and $n(d)$ would have been the optimal sample size. Note that $n(d)$ depends on d, α and γ, where γ depends, in turn on the unknown F. It may be noted that by virtue of (4.5.16),

$$\hat{\theta}_{U,N(d)-1} - \hat{\theta}_{L,N(d)-1} > 2d \geq \hat{\theta}_{U,N(d)} - \hat{\theta}_{L,N(d)}, \qquad d > 0, \tag{4.5.19}$$

so that by virtue of (4.5.17), (4.5.18) and (4.5.18), we have

$$N(d)/n(d) \to 1 \quad \text{a.s.,} \quad \text{as } d \downarrow 0. \tag{4.5.20}$$

The same linearity result also ensures that as $n \to \infty$,

$$n^{1/2}(\hat{\theta}_{n(R)} - \tfrac{1}{2}(\hat{\theta}_{L,n} + \hat{\theta}_{U,n})) \to 0 \quad \text{a.s.,} \tag{4.5.21}$$

so that by (4.3.7), (4.5.20), (4.5.21) and some standard arguments it follows that (4.5.6) holds. To verify (4.5.12) in this case, one needs to replace (4.5.17) by the following: For every $\varepsilon > 0$ and some positive s (>1), there exist an n_ε and K_s ($>\infty$), such that

$$P\{|\gamma n^{1/2}(\hat{\theta}_{U,n} - \hat{\theta}_{L,n}) - 2A\tau_{\alpha/2}| > \varepsilon\} \leq K_s n^{-s} \quad \forall n \geq n_\varepsilon. \tag{4.5.22}$$

Indeed, under (4.3.5), (4.5.22) holds—again, rates of convergence of $\sup \{n^{-1/2} |S_n(\theta + b) - S_n(\theta) + nb\gamma| : b \in [-cn^{-1/2} \log n, cn^{-1/2} \log n]\}$ (to 0), studied by Sen and Ghosh (1971) and Sen (1980a) provides the proof. The basic approach to this problem is due to Jurečková (1969) (who, however, considered only the convergence in probability, without any specific rate). By using the monotonicity of $S_n(b)$ in b, one may replace the "sup" by that of a "max" over a given number of discrete values of b (plus a remainder term bounded by some given $\varepsilon > 0$), and at each b, one may then use some precise order of convergence of the empirical distributions to the true ones, which provide the desired rate of convergence. For some of these details, we may refer to the Appendix, § A.4 of Sen (1981d).

This method of inversion of some test statistics to obtain confidence intervals for the related parameter works out well for other problems as well. For the difference of location parameters, in the two-sample model, one may use the Wilcoxon–Mann–Whitney or other two-sample statistics (with monotone scores) and proceed on parallel lines. This two-sample model is a particular case of a simple regression model. For this simple regression model, Ghosh and Sen (1971) used the Kendall tau-statistic to obtain a bounded length (sequential) confidence interval for the regression slope. Later on, Ghosh and Sen (1972) considered a general class of linear rank statistics and used them to obtain bounded width (sequential) confidence intervals for the regression parameter. In each case, (4.5.6) and (4.5.12) hold under appropriate regularity conditions; (4.3.5) again suffices for the case of linear rank statistics.

In the case of M-statistics, this Type B sequential procedure works out with some minor adjustment. First, unlike the case of rank statistics, the M-statistics are not, generally, distribution-free (under suitable null hypotheses), so that if, as in before (4.3.8), we want to find out the percentile points of the null distribution, we may have difficulties. However, defining s_n^2 as in (4.3.16), we may take these as $\pm \tau_{\alpha/2} s_n (\sqrt{n})$, and then, as in before (4.3.17), we may define $\hat{\theta}_{n,U}$ and $\hat{\theta}_{n,L}$ and with these, we may proceed as in (4.5.16). Here also, the asymptotic linearity results (such as in (4.3.20) of Jurečková and Sen (1981a, b) provide the necessary tools.

In the case of L-estimators, one essentially has a Type A confidence region based on the point estimator in (4.3.21) and the estimator of ν^2 in (4.3.22).

The theory has also been extended to the multivariate case; see Sen (1981d, Chap. 10) and Sen (1984f). The coordinatewise linearity results and the rates of convergence of the sample counterpart of the dispersion matrices provide the necessary tools.

We conclude this section with some remarks on the asymptotic normality of the stopping times $N(d)$ in (4.5.3) and (4.5.16). By virtue of the inequality, following (4.5.7), and the definition of $n(d)$, $d > 0$, the asymptotic normality of $(n(d))^{-1/2}(N(d) - n(d))$ (as $d \downarrow 0$) is essentially tied down to that of $n^{1/2}(s_n^2 - \sigma_\theta^2)$, for Type A procedures. As such, this holds under very general conditions. On the other hand, for Type B procedures, by (4.5.19), one needs to study the asymptotic normality of

$$n^{1/2}\{\gamma n^{1/2}(\hat{\theta}_{U,n} - \hat{\theta}_{L,n}) - 2A\tau_{\alpha/2}\} \tag{4.5.23}$$

for this purpose. This again calls for "second order linearity" results on rank statistics. Though some progress on this line has been made by Hušková (1982) and Hušková and Jurečková (1981), (1984), there is room for further developments.

4.6. Some general remarks. The AMRE property in (4.2.9) on the asymptotic efficiency result in (4.5.12) may be considered as a "first order" result only. In (4.2.9), we have by (4.2.5) and (4.2.8),

$$\lambda_F^*(c) = \lambda_F^0(c) + o(c^{q/(q+1)}), \quad \text{as } c \downarrow 0, \tag{4.6.1}$$

and there remains a lot of room for further study of the actual order of excess of $\lambda_F^*(c)$ over $\lambda_F^0(c)$. Similarly, if we define the *regret function*

$$\pi(d) = EN(d) - n(d), \tag{4.6.2}$$

then by (4.5.12), for $\psi(n) \equiv n^{1/2}$, $\pi(d) = o(d^{-2})$, as $d \downarrow 0$. Here also, there is ample scope for the study of the actual order of $\pi(d)$, as $d \downarrow 0$. In this context, the asymptotic normality result on $(n(d))^{-1/2}(N(d) - n(d))$ (as $d \downarrow 0$) provides some information, though it fails to provide suitable bounds on $\pi(d)$. In the point estimation problem, we have additionally

$$\pi^*(c) = \lambda_F^*(c) - \lambda_F^0(c) = c\{EN_c - n_c^0\} + (n_c^0)^{-q}\{(n_c^0)^q \rho_{N_c^0}(F) - (n_v^0)^q E g(|T_{N_c} - \theta|)\}, \tag{4.6.3}$$

so that one needs to study more precisely some asymptotic expansions for $(n_c^0)^q Eg(|T_{N_c}-\theta|)$ along with the rate of convergence of $(n_c^0)^q \rho_{n^0}(F)$ to $\rho^*(F)$ in (4.2.3). When $g(t) \equiv t^2$ and T_n is some linear statistic, some asymptotic expansion for $\pi(d)$ in (4.6.2) and $\pi^*(c)$ in (4.6.3) have been obtained by using some nonlinear renewal theory by Woodroofe (1977), (1982) and Lai and Siegmund (1977), (1979), among others. However, for the majority of nonparametric estimators, both in the point and interval estimation problems, the regularity conditions needed to verify the applicability of these theorems, pose serious computational problems. There is a profound need for further refinements of such nonlinear renewal theorems for general (nonlinear) nonparametric statistics, and we like to list this as an open area for future research.

In the point estimation problem (AMRE), we have discussed the use of M- and L-estimators for which the score function ψ is constant outside a compact interval and $\phi(u)$ is 0 on $[0, \varepsilon] \cup [1-\varepsilon, 1]$, for some $0 < \varepsilon < \frac{1}{2}$. Though this may be justifiable on the ground of robustness and elimination of the devastating effects of outliers, from the theoretical point of view, there may be a genuine question whether such boundedness conditions may be replaced by suitable "growth" conditions on ϕ or ψ. This may indeed be done, but would naturally result in the existence of higher order moments of the underlying distribution. The method of attack of Jurečková and Sen (1982) may, however, have to be replaced by more direct approaches for verifying the needed uniform integrability conditions—and this may require more stringent regularity conditions. A more serious problem may arise in the case of regression models, where, even for bounded ψ or ϕ, the method of attack of Jurečková and Sen (1982) may not work out. Further work on this line is under investigation.

In the parametric model, Mukhopadhyay (1980) has considered a modification of the Stein two-stage procedure (where the initial sample size is made to depend on c or d) for which the AMRE (or the asymptotic efficiency) property holds; however, this may not have the higher order asymptotic efficiency properties. The basic result of Mukhopadhyay (1980) extends readily to the general class of nonparametric estimators considered here. However, these do not require any less stringent regularity conditions. Moreover, the idea of updating the variance estimator to enhance the efficiency (underlying a sequential procedure) is so natural that one would be naturally inclined to use a sequential procedure instead of a two-stage procedure, along with the hope that it would be more efficient (at least in the "higher order" sense). This is indeed possible to verify this in various specific cases and a general theory towards this effect is under investigation.

We conclude this chapter with some discussion on the use of *adaptive* score functions for nonparametric statistics in the context of estimation of location or regression parameters. First, for the M-estimators of location (or regression), we note that the M-statistics are not generally *scale-equivariant* (so that the derived estimators may also fail to be scale-equivariant). Based on the concept of *regression* (or *location*) *quantiles* (due to Koenker and Bassett (1978) and elaborated in Ruppert and Carroll (1980)), an adaptive, scale-equivariant version of M-estimators in linear models has been considered in Jurečková and Sen (1984), and, in the conventional nonsequential case, various (asymptotic) properties of such adaptive M-estimators have been studied by them. Generalizations of their theory to the sequential case is under investigation. Secondly, both the M-estimators and R-estimators rest on the use of a prescribed score function, and, ideally, one intends to choose such a score function in such a way that the estimators have some (asymptotic) optimality properties. However, such an optimal choice of the score function depends on the form of the underlying d.f. F (generally, unknown), and, hence, some adaptive procedures are usually prescribed to achieve this goal. Martinsek (1984) has treated this problem in some generality. Sequentially adaptive asymptotically efficient rank estimators have also been recently studied by Hušková and Sen (1984). In their proposal, use of *Legendre polynomials* (in a *Fourier representation* of the score function) has led to some well defined stopping rules leading to asymptotically efficient, adaptive R-estimators of location and regression. Further generalizations to more appropriate sequential problems are under consideration.

CHAPTER 5

Nonparametric Repeated Significance Tests

5.1. Introduction. For parametric or nonparametric tests of statistical hypotheses to have good power properties (against alternatives not too far away from the null ones), generally, one needs to base such tests on samples of reasonably large sizes. In some situations, these sample observations can be drawn all at the same time, so that a fixed sample size procedure may be constructed. However, in general, in clinical trials or other medical experimentations, the sample units may not all be available simultaneously; they may enter into the scheme either in (relatively small) batches or even sporadically over time. In such a case, though it may be customary to set, in advance, a *target sample size*, say N, usually large, on which to base a test, a complete collection of data may involve a considerable *waiting time*, and one may therefore like to have *interim analyses* either at regular intervals of time or at regular subsample sizes. In the extreme case, one may also consider a *statistical monitoring* from the beginning, whereby a statistical testing scheme is adopted either continuously over time or at the successive sample sizes $\{n \leq N\}$. In such a case, one has the flexibility to terminate the study at an intermediate stage, contingent on the accumulating statistical evidence.

We have considered *progressive censoring schemes* (PCS) in Chapter 2, where basically one has a follow-up study: The *responses* occur sequentially over time, while the entries may be simultaneous or staggered. On the contrary, we are primarily concerned here with *staggered entry plans* where the responses are immediate and do not entail a follow-up scheme. A more complicated scheme arises in a PCS under staggered entry plans (viz. § 2.6). A common feature of all these procedures is the provision of making *repeated statistical tests* or interim statistical analysis on accumulating data with a view to making an early termination of the study whenever a clear statistical picture is available. From the point of view of *medical ethics*, such a RST (repeated significance testing) procedure may serve a very useful purpose: If, at any early stage $n \ (\leq N)$, a clear statistical decision can be reached, the trial may be terminated at that stage, leading to possible savings of time of experimentation and lives of experimental units as well. The statistical evidence acquired from a terminated trial may also facilitate the immediate implementation of corrective medical actions on oncoming subjects. From the statistical point of view, however, RST may entail complications: Repeated testing on accumulating (dependent) data may unduly increase the risk of making incorrect decisions, and hence, sufficient care must be exercised to incorporate proper statistical methodology ensuring prescribed margin of errors. This methodology, though related, is not the same as in the case of PCS, and, in this chapter, we like to discuss this in detail.

At this point, we may pressure some distinctions between RST, TST (truncated sequential testing) and GST (group sequential testing) procedures. In a RST, for each n ($\leq N$), one may use the fixed sample size (optimal or desirable) test statistic T_n, and the testing procedure is then based on the partial sequence $\{T_n; n \leq N\}$ and conducted in a *quasi-sequential manner*. Thus, a *stopping number* K ($\leq N$) is properly defined in terms of the $T_n, n \leq N$, and this dictates the *stopping rule* too. In a TST, one uses the probability (or likelihood) ratio or some other criterion leading to a sequence $\{T_n^*, n \leq N\}$ of test statistics, along with a stopping number K^* ($\leq N$), where by construction $P\{K^* \leq N\} = 1$. Though $\{T_n\}$ and $\{T_n^*\}$ may be related, they may be normalized differently, leading to different procedures. For example, in the case of sampling from a normal distribution (with mean μ and variance σ^2), for testing $H_0: \mu = 0$ vs. $H_1: \mu = \mu_1 > 0$, in a RST, one has

$$T_n = \sqrt{n}\,\bar{X}_n/s_n \quad \text{where} \quad \bar{X}_n = \frac{1}{n}\sum_{i=1}^{n} X_i \quad \text{and} \quad s_n^2 = \frac{1}{n-1}\sum_{i=1}^{n}(X_i - X_n)^2,$$

$$(5.1.1)$$

for $n \geq 2$, while in a TST, one has

$$T_n^* = n(\mu_1 - \mu_0)(\bar{X}_n - \tfrac{1}{2}(\mu_1 + \mu_0))/s_n, \qquad n \geq 2, \quad \mu_0 = 0. \qquad (5.1.2)$$

Thus, if one uses $\max\{T_k : k \leq n\}$ (or $\max\{T_k^* : k \leq n\}$) or their two-sided versions, the testing procedures are different for the two schemes. In a GST, of each stage, a subsample of size m (> 1) is considered, and statistics $\{Z_k, k \geq 1\}$ based on these subsamples are incorporated in the construction of the sequences $\{T_n^0, n \leq N\}$ on which a (quasi-) sequential test is based. These T_n^0 are generally different from the T_n or T_n^*.

For some simple parametric models, a nice account of some RST is given in Armitage (1975), where other references are cited too. The inflation of the overall significance level due to multiple testing on the accumulating data has been clearly depicted there for some simple procedures: The suggestion is to use smaller significance levels for the individual tests so that the overall test has a prescribed margin of Type I error. This feature is shared by the RST in the nonparametric cases as well. Fortunately, in this context, some *invariance principles* for $\{T_n; n \leq N\}$ may be used with advantage to provide asymptotic solutions which remain applicable to both parametric and nonparametric situations. For various nonparametric statistics, these invariance principles have been systematically presented in (Sen (1981b, Part I)). Here, we shall take a pragmatic view of these basic results and incorporate them in the general formulation of nonparametric RST procedures.

Section 5.2 is devoted to a unified treatment of invariance principles for nonparametric statistics through a martingale characterization of *locally most powerful invariant test* statistics (viz. Sen (1981a)). The implications of the *contiguity of probability measures* play a vital role in this context. These invariance principles are incorporated in § 5.3 in the formulation of the general (asymptotic) theory of nonparametric repeated significance tests. Section 5.4 is devoted to repeated multiple comparisons tests. The concluding section deals with some general remarks and observations.

5.2. Invariance principles for nonparametric statistics. In a genuine nonparametric hypothesis testing problem, the null hypothesis H_0 relates to the invariance of the joint distribution of the sample observations under some (finite) group ($\mathcal{G}$) of transformations which maps the sample space ($\mathcal{E}$) onto itself. This group $\mathcal{G}$, applied to a sample point E, defines an *orbit* $\mathcal{E}^* = \{gE : g \in \mathcal{G}\}$, and under H_0, the (conditional) distribution of E on $\mathcal{E}^*$ is uniform; all the (discrete) mass points in E^* are conditionally equally likely. Thus, under H_0, it may be possible to induce a transformation

$$E \rightarrow (T, Z) \tag{5.2.1}$$

where T (possibly vector valued) is the *minimal sufficient statistic* (MSS) and Z (possibly, vector valued) is a (maximal) statistic independent of T; Z is termed a *maximal statistical noise* (MSN). In the context of *invariant tests*, the concept of MSN coincides with that of *maximal invariants*. Thus, for T, a MSS, the conditional distribution of Z, given $T = t$, is independent of t (and P_θ^t), i.e., Z is distribution-free under H_0. Hence, a test function $\phi(E)$, depending on E only through Z, i.e., $\phi(E) \equiv \phi(Z)$, is distribution-free (under H_0). This provides a basic characterization of nonparametric tests.

Consider first the hypothesis of *randomness*. Let $X_1, \ldots, X_n$ be n independent r.v. with continuous d.f. $F_1, \ldots, F_n$, respectively, all defined on the real line $E = (-\infty, \infty)$. Consider the null hypothesis $H_0 : F_1 = \cdots = F_n = F$ (unknown). Under H_0, the joint distribution of $X_1, \ldots, X_n$ remains invariant under any permutation of the arguments. Let $\mathbf{T} = \{X_{n:1} < \cdots < X_{n:n}\}$ be the vector of *order statistics* and let $\mathbf{R}_n = (R_{n1}, \ldots, R_{nn})$ be the vector of *ranks* (i.e., $X_i = X_{n:R_{ni}}, 1 \le i \le n$); ties neglected, with probability one, by virtue of the assumed continuity of the F_i. Here $\mathbf{T}$ is the MSS and $\mathbf{R}_n$ is the MSN (maximal invariant); under H_0, $\mathbf{R}_n$ takes on each permutation of $(1, \ldots, n)$ with the common probability $(n!)^{-1}$. Hence, a test depending on the X_i through $\mathbf{R}_n$ alone is distribution-free.

The choice of such a test statistic is naturally influenced by the type of alternative hypotheses one has in mind. Consider the *two-sample problem* where for some $n_1 : 1 \le n_1 < n$ ($n_2 = n - n_1$), $F_1 = \cdots = F_{n_1} = F$ and $F_{n_1+1} = \cdots = F_n = G$, so that H_0 reduces to $F = G$, while we may be interested in the alternatives $F \ne G$ arising due to difference in locations/scale parameters or other forms. The *simple regression model* also pertains to this hypothesis of randomness: The model relates to

$$F_i(x) = F(x - \beta c_i), \qquad x \in E, \quad i = 1, \ldots, n, \tag{5.2.2}$$

where the c_i are known constants, not all equal, and β is an unknown parameter. Under H_0, $\beta = 0$. The two-sample location model is a special case of (5.2.2), where one may take $c_1 = \cdots = c_{n_1} = 0, c_{n_1+1} = \cdots = c_n = 1$. For the model (5.2.2), one may consider a rank statistic of the form

$$T_n = \sum_{i=1}^{n} (c_i - \bar{c}_n) a_n(R_{ni}), \qquad (5.2.3)$$

where $\bar{c}_n = n^{-1} \sum_{i=1}^{n} c_i$ and $a_n(1), \ldots, a_n(n)$ are suitable *scores*. T_n is termed a *linear rank statistic*. Typically, one chooses

$$a_n(k) = E\phi(U_{nk}) \quad \text{for } k = 1, \ldots, n, \qquad (5.2.4)$$

where $\phi = \{\phi(t), t \in (0, 1)\}$ is a suitable (integrable) *score generating function* and $U_{n1} < \cdots < U_{nn}$ are the ordered r.v. of size n from the uniform $(0, 1)$ d.f. For a given family of underlying F and for a specific type of alternative hypotheses, one may choose ϕ in such a way that the corresponding T_n in (5.2.3) is a *locally most powerful rank* (LMPR) test statistic. For example, for (5.2.2), if we let $\phi(u) = -f'(F^{-1}(u))/f(F^{-1}(u))$, $0 < u < 1$, then the corresponding T_n is a LMPR test statistic for $H_0: \beta = 0$ vs. $H_1: \beta > 0$. Actually, if $U_1, \ldots, U_n$ are i.i.d.r.v. with the uniform $(0, 1)$ d.f., and if we let $U_i = F(X_i)$, $i \geq 1$ and

$$T_n^* = \sum_{i=1}^{n} (c_i - \bar{c}_n) \phi(U_i), \qquad (5.2.5)$$

then by (5.2.3)–(5.2.5), we have $E(T_n^* | \mathbf{R}_n, H_0) = T_n$, $\forall n \geq 1$. Let us assume that ϕ is square integrable inside $(0, 1)$ and denote by $\bar{\phi} = \int_0^1 \phi(u) \, du$ and $A^2 = \int_0^1 \phi^2(u) \, du - \bar{\sigma}^2$. Then, $ET_n^* = 0$ and $E(T_n^{*2}) = C_n^2 A^2$ where $C_n^2 = \sum_{i=1}^{n} (c_i - \bar{c}_n)^2$. Also, $E(T_n | H_0) = 0$ and $E(T_n^2 | H_0) = C_n^2 A_n^2$, where $A_n^2 = (n-1)^{-1} \sum_{i=1}^{n} [a_n(i) - \bar{a}_n]^2$, $\bar{a}_n = n^{-1} \sum_{i=1}^{n} a_n(i)$. Further, $E[(T_n^* - T_n)^2 | H_0] = E(T_n^{*2}) - E(T_n^2 | H_0) = C_n^2(A^2 - A_n^2)$, $n \geq 2$. As such, it is easy to verify that

$$E\{(T_n - T_n)^2 | H_0\}/E(T_n^2) = A^2/A_n^2 - 1 \to 0 \quad \text{as } n \to \infty. \qquad (5.2.6)$$

Thus, $C_n^{-1}(T_n - T_n^*) \to 0$, in probability, under H_0, as $n \to \infty$. On the other hand, T_n^* involves independent summands, and hence, whenever $\max \{(c_i - \bar{c}_n)^2/C_n^2 : 1 \leq i \leq n\} \to 0$ and $0 < A^2 < \infty$, $C_n^{-1} T_n^*$ has asymptotically the normal distribution with 0 mean and unit variance. Thus, under H_0 and the same regularity conditions, as $n \to \infty$,

$$C_n^{-1} T_n \underset{\mathcal{D}}{\to} \mathcal{N}(0, A^2). \qquad (5.2.7)$$

The same proof applies to the asymptotic multinormality of C_n^{-1} $(T_{n_1}, \ldots, T_{n_m})$ for any (fixed) m (≥ 1) and $\{n_1, \ldots, n_m\}$, such that $C_{n_j}^2/C_n^2 \to t_j$, $1 \leq j \leq m$, where $0 \leq t_1 < \cdots < t_m \leq 1$. Further, note that by (5.2.4), for every n (≥ 1) and k $(1 \leq k \leq n)$,

$$(n+1)^{-1} k a_{n+1}(k+1) + (n+1)^{-1}(n+1-k) a_{n+1}(i) = a_n(k), \qquad (5.2.8)$$

so that by (5.2.3) and some standard arguments (viz. Sen and Ghosh (1972)),

$$E\{T_{n+1} | \mathcal{B}(\mathbf{R}_n), H_0\} = T_n \quad \forall n \geq 1. \qquad (5.2.9)$$

This martingale property ensures *tightness* of suitable processes constructed from $\{T_1, \ldots, T_n\}$. Hence, referring back to Sen (1981d, Chap. 4), we may present the following invariance principle.

For every n (≥ 1), define $n(t) = \max\{k : C_k^2 \leq tC_n^2\}$, $0 \leq t \leq 1$ and let $Y_n(t) = T_{n(t)}/\{C_n A_n\}$, $0 \leq t \leq 1$, $Y_n = \{Y_n(t), 0 \leq t \leq 1\}$. Let $W = \{W(t), 0 \leq t \leq 1\}$ be a standard Wiener process on $[0, 1]$. Then, for square integrable ϕ and the c_i satisfying the Noether condition, mentioned before,

$$Y_n \underset{\mathcal{D}}{\rightarrow} W, \quad \text{in the Skorokhod } J_1\text{-topology on } D[0, 1]. \tag{5.2.10}$$

(The result is true for the entire interval $[0, \infty)$, with an extended topology, and further under d_q-metric too, where q is continuous, $\nearrow$ and $\int q^{-2}(t)\, dt < \infty$.)

We may note that the scores need not be defined by (5.2.4). The result (5.2.10) remains true, for any other set $\{a_n^*(1), \ldots, a_n^*(n)\}$ of scores, whenever $\max\{n^{-1/2} \mid a_n(i) - a_n^*(i) \mid : 1 \leq i \leq n\} \rightarrow 0$, as $n \rightarrow \infty$. We may also remark that $\phi(\cdot)$ in (5.2.5) may be interpreted as the derivating of the log-density function of the X_i (with respect to some underlying distribution), so that T_n^* is a LMP test statistic and T_n is the corresponding LMPR test statistic. As such, the martingale property of T_n (or T_n^*) also follows from the martingale property of LMPR (or LMP) test statistics, and invariance principles for such LMPR test statistics, developed in Sen (1981a) may be used to provide an alternative proof for (5.2.10).

None of these approaches remains valid in the case when the null hypothesis of invariance is not true—the basic martingale structure may not hold. Though, in a general setup, the treatment may be quite involved, for *contiguous alternatives* (as in Chapters 2 and 3), one may have a very simple treatment. Note that the tightness of $\{Y_n\}$ under H_0 and the contiguity of the probability measures (under the sequence of alternatives to that under H_0) ensure the tightness of $\{Y_n\}$ under the alternative as well. Further, for such contiguous alternatives, for the convergence of finite dimensional distributions (f.d.d) of $\{Y_n\}$, one may again use the stochastic convergence of $C_n^{-1}(T_n - T_n^*)$ (to 0, under H_0) and contiguity to conclude that the same stochastic convergence holds under such an alternative as well. Finally, for $C_n^{-1}T_n^*$ one may use the standard central limit theorems and conclude that $C_n^{-1}T_n^*/A_n$ converges in distribution to a normal one with a finite mean and unit variance—a similar result holds for $C_n^{-1}A_n^{-1}(T_{n_1}^*, \ldots, T_{n_m}^*)$. Hence, under such a sequence of contiguous alternatives, Y_n converges weakly to a drifted Wiener process $(W + \xi)$, where $\xi = \{\xi(t), 0 \leq t \leq 1\}$ is a drift function on $[0, 1]$. For various specific types of (viz. location, scale or regression) alternatives, the form of ξ has been studied in detail in Sen (1981d, Chap. 4).

Consider next the hypothesis of *sign invariance*. Let $X_1, \ldots, X_n$ be n independent r.v. with a common continuous d.f. F, defined on $E = (-\infty, \infty)$. The null hypothesis (H_0) relates to the symmetry of F around a specified median (say, 0), so that

$$H_0 : F(x) + F(-x) = 1 \quad \forall x \in E. \tag{5.2.11}$$

In testing H_0, one may have shift alternatives in mind (i.e., $F(x) = F_0(x - a)$, F_0

symmetric and $a \neq 0$) or a departure from symmetry in some other form. For this model, one may consider a *signed-rank statistic* of the form

$$S_n = \sum_{i=1}^{n} \text{sign } X_i a_n^*(R_{ni}^+), \qquad (5.2.12)$$

where the scores $a_n^*(k)$ are defined as in (5.2.4) with a suitable score generating function ϕ^*, and $R_{ni}^+ = $ rank of $|X_i|$ among $|X_1|, \ldots, |X_n|$, for $i = 1, \ldots, n$. Typically, $\phi^*(u) = \phi((1+u)/2)$, $0 < u < 1$ where ϕ is a skew-symmetric function (i.e., $\phi(u) + \phi(1-u) = 0$, $\forall u \in (0, 1)$). Let $\mathbf{T}^+ = \{X_{n:1}^+ < \cdots < X_{n:n}^+\}$ be the vector of order statistics of the $|X_i|$, $1 \leq i \leq n$, $\mathbf{R}_n^+ = \{R_{n1}^+, \ldots, R_{nn}^+\}$ and $\mathbf{S}_n = (\text{sign } X_1, \ldots, \text{sign } X_n)$. Here T^+ is the MSS and $(\mathbf{R}_n^+, \mathbf{S}_n)$ is the MSN; actually, under H_0, $\mathbf{R}_n^+$ and $\mathbf{S}_n$ are stochastically independent, where $\mathbf{R}_n^+$ takes on each permutation of $(1, \ldots, n)$ with the actual probability $(n!)^{-1}$, while $\mathbf{S}_n$ takes on each of the 2^n sign inversions $(+1, \ldots, \pm 1)$ with the common probability 2^{-n}. Thus, S_n in (5.2.12) is genuinely distribution-free under H_0. If we let $S_n^* = \sum_{i=1}^{n} \phi(F(X_i))$ $(= \sum_{i=1}^{n} \text{sign } X_i \phi^*(F^*(|X_i|)))$ (where $F^*(x) = F(x) - F(-x)$), then we have $E(S_n^* \mid \mathbf{R}_n^+, \mathbf{S}_n, H_0) = S_n$, $\forall n \geq 1$. Also, note that for square integrable and skew-symmetric ϕ, $\bar{\phi} = \int_0^1 \phi(u) \, du = 0$ and $A^2 = \int_0^1 \phi^2(u) \, du = \int_0^1 [\phi^*(u)]^2 \, du < \infty$, so that if we let $A_n^{*2} = (1/n)\sum_{i=1}^{n} [a_n^*(i)]^2$, then

$$n^{-1}E[(S_n^* - S_n)^2 \mid H_0] = n^{-1}\{E(S_n^{*2} \mid H_0) - E(S_n^2 \mid H_0)\}$$

$$= n^{-1}\{nA^2 - nA_n^{*2}\} = A^2 - A_n^{*2} \to 0 \quad \text{as } n \to \infty, \quad (5.2.13)$$

so that under H_0, $n^{-1/2}(S_n^* - S_n) \to 0$, in probability, as $n \to \infty$. On the other hand, for S_n involving i.i.d. summands, the classical control limit theorem applies whenever $0 < A < \infty$. Hence, under H_0 and $0 < A < \infty$, $n^{-1/2}S_n \underset{\mathscr{D}}{\to} \mathcal{N}(0, A^2)$. In a similar manner, one can also obtain the asymptotic multinormality of $n^{-1/2} (S_{n_1}, \ldots, S_{n_m})$ for every (fixed) m (≥ 1), whenever $n_j/n \to t_j$: $0 \leq t_1 < \cdots < t_m \leq 1$. By the use of (5.2.8) and the stochastic independence of $(\mathbf{R}_n^+, \mathbf{S}_n)$, Sen and Ghosh (1971) have shown that under H_0,

$$E(S_{n+1} \mid \mathbf{R}_n^+, \mathbf{S}_n, H_0) = S_n \quad \forall n \geq 1. \qquad (5.2.14)$$

This martingale property has again been exploited in the proof of the tightness of suitable stochastic processes constructed from $\{S_1, \ldots, S_n\}$; we may refer to Sen (1981b, Chap. 5) for the details and state the following basic result here.

For every n, let $n(t) = [nt]$ be the largest integer $\leq nt$, for $0 \leq t \leq 1$, and let $Y_n^0 = \{Y_n^0(t) = n^{-1/2}S_{n(t)}, 0 \leq t \leq 1\}$. Then, under H_0, as $n \to \infty$,

$$Y_n^0 \underset{\mathscr{D}}{\to} W, \quad \text{in the } J_1\text{-topology on } D[0, 1]. \qquad (5.2.15)$$

Here also, one may consider other scores $\{l_n^*(1), \ldots, l_n^*(n)\}$, such that $n^{-1/2}(\max \{|a_n^*(i) - l_n^*(i)| : 1 \leq i \leq n\}) \to 0$, then (5.2.15) holds for such signed-rank statistics (where in (5.2.12) the $a_n^*(k)$ are replaced by the $l_n^*(k)$). Further, Y_n^0 may be extended to $[0, \infty)$ and (5.2.15) may be established with an extended topology and in the d_q-metric as well. Finally, here also S_n can be characterized as a LMPR test statistic (against suitable alternatives), so that the martingale property in (5.2.11) may also be established by using the martingale structure of LMPR test statistics, developed by Sen (1981a) may also be used to provide an alternative proof of (5.2.15).

For contiguous alternatives, Y_n^0 converges weakly to a drifted Wiener process $W + \xi^0$, where $\xi^0 = \{\xi^0(t), 0 \le t \le 1\}$ is the drift function and it depends on the specific alternative sequence. For a fixed alternative, some almost sure invariance principles for such signed rank statistics have also been discussed in Sen (1981d, Chap. 5).

Next, we consider the hypothesis of *matching invariance* (or *bivariate independence*). Let (X_i, Y_i), $i = 1, \ldots, n$ be n i.i.d.r.v.'s with a continuous (bivariate) d.f. F, defined on E^2. Consider the null hypothesis

$$H_0 : F(x, y) \equiv F(x, \infty) F(\infty, y) \quad \forall (x, y) \in E^2, \tag{5.2.16}$$

against suitable alternatives (viz. positive or negative association, etc.). Let R_{ni} (and S_{ni}) be the rank of X_i (and Y_i) among $X_1, \ldots, X_n$ (and $Y_1, \ldots, Y_n$), for $i = 1, \ldots, n$. As in (5.2.4), consider two sets of scores $a_n(1), \ldots, a_n(n)$ and $b_n(1), \ldots, b_n(n)$, generated by two score-generating functions ϕ_1 and ϕ_2, respectively. Then, a typical rank test statistic is of the form

$$M_n = \sum_{i=1}^{n} [a_n(R_{ni}) - \bar{a}_n][b_n(S_{ni}) - \bar{b}_n] \tag{5.2.17}$$

where $\bar{a}_n = n^{-1} \sum_{i=1}^{n} a_n(i)$ and $\bar{b}_n = n^{-1} \sum_{i=1}^{n} b_n(i)$. Under H_0 in (5.2.16), the two stochastic vectors $\mathbf{R}_n = (R_{n1}, \ldots, R_{nn})'$ and $\mathbf{S}_n = (S_{n1}, \ldots, S_{nn})'$ are mutually independent, and each takes on each permutation of $(1, \ldots, n)$ with the common probability $(n!)^{-1}$. Hence, M_n is distribution-free under H_0. If we define $M_n^* = \sum_{i=1}^{n} \phi_1(F_1(X_i))\phi_2(F_2(Y_i))$ where $F_1(x) = F(x, \infty)$ and $F_2(y) = F(\infty, y)$, then letting $\phi_j = \int_0^1 \phi_j(u)\, du = 0$ for $j = 1, 2$, it follows that $M_n = E(M_n^* \mid \mathbf{R}_n, \mathbf{S}_n, H_0)$, and $n^{-1} E_0(M_n^* - M_n)^2 \to 0$ as $n \to \infty$. Consequently, by the same technique as in the case of linear rank statistics, we have an invariance principle for the M_k, $k \le n$, where again the martingale property of $\{M_n; n \ge 1\}$, under H_0, plays a vital role. For contiguous alternatives too, this invariance principle extends naturally, and the details are worked out in Sen (1981d, Chap. 6).

Most of these nonparametric statistics may also be characterized as locally most power invariant (LMPI) test statistics (with respect to appropriate (parametric) forms of alternatives). As such, a martingale characterization of such rank statistics may also be established through the martingale property of LMP test statistics and their conditional expectations (given a nondecreasing sequence of sub-sigma fields), under the usual regularity conditions; these are worked out in Sen (1981a). The mean square equivalence result (in (5.2.6) or (5.2.13)) in the specific case provides an easy access to verify the regularity conditions for the usual invariance principles for martingales.

U-statistics, as have been introduced in Chapters 3 and 4, also play a vital role in nonparametrics. Since the U-statistics form a reversed martingale sequence, invariance principles for the latter work out well for the former too.

Here also, the Hoeffding (1948) projection (or Hoeffding (1961) decomposition) provides a great simplification (viz. Miller and Sen (1972)), and most of these details are worked out in (1981d, Chap. 3).

Linear combinations of functions of order statistics play a vital role in robust statistical inference. Often, these are linear combinations of functions of selected sample quantiles. For sample quantiles, Bahadur (1966) has considered an almost sure representation (in terms of an average of i.i.d.r.v.'s) and this provides an easy access to establish invariance principles (both weak and strong ones) under very general regularity conditions; these are discussed in detail in Sen (1981d, § 7.3). In the other situation, typically, we have a statistic of the form

$$T_n = \sum_{i=1}^{n} c_{n,i} g(X_{n:i}) = \int_{-\infty}^{\infty} \phi_n(F_n(x)) g(x) \, dF_n(x) \qquad (5.2.18)$$

where $X_{n:1} \leq \cdots \leq X_{n:n}$ are the ordered r.v.'s, F_n is the sample d.f. and $c_{n,i} = n^{-1} \phi_n(i/(n+1)) = n^{-1} \phi_n(t)$, $(i-1)/n < t \leq i/n$, $i = 1, \ldots, n$, and where $\phi_n(\cdot)$ converges (a.e.) to a smooth function $\phi(\cdot)$, as $n \to \infty$. Various authors have employed diverse techniques for the study of the asymptotic theory of T_n. In Sen (1981d, § 7.4), it has been shown that under quite general regularity condition (on $g(\cdot)$ and $\phi_n(\cdot)$), $\{T_n\}$ may be approximated by a reverse martingale sequence, and this approximation paves the way for the weak as well as strong invariance principles for the T_n. Unlike the case of rank statistics, T_n is not a distribution-free statistic and its (asymptotic) variance function is a functional of the underlying d.f. Strongly consistent estimators of this functional and their asymptotic properties were studied by Gardiner and Sen (1979); see also Sen (1981d, § 7.6), (1984c). The latter paper deals with the role of jackknifing of L-estimators and asymptotic properties of jackknifed variance estimators of T_n.

Consider the linear model: $X_i = \boldsymbol{\beta}' \mathbf{c}_i + e_i$, $i \geq 1$, where the e_i are i.i.d.r.v. with the d.f. F defined on E $(=(-\infty, \infty))$, the $\mathbf{c}_i$ (vectors) are known, while $\boldsymbol{\beta}$ is a vector of unknown parameters. The *least squares* estimators of $\boldsymbol{\beta}$ (i.e., $\hat{\boldsymbol{\beta}}_n = (\sum_{i=1}^{n} \mathbf{c}_i \mathbf{c}_i')^{-1} \sum_{i=1}^{n} \mathbf{c}_i X_i$) are linear functions of the X_i, and hence, the invariance principles can be easily obtained when $Ee_i^2 < \infty$ and the $\mathbf{c}_i$ satisfy the (generalized) Noether condition: $\max_{1 \leq k \leq n} \{\mathbf{c}_k' (\sum_{i=1}^{n} \mathbf{c}_i \mathbf{c}_i')^{-1} \mathbf{c}_k : 1 \leq k \leq n\} \to 0$, as $n \to \infty$. Among other estimators, a general class (including the least squares and maximum likelihood estimators as special cases) deserves special mention: This is the class of so-called M-estimators. Typically, one considers a *score-function* $\psi = \{\psi(x), x \in E\}$ (a nondecreasing, skew-symmetric function) and considers the vector process

$$\mathbf{S}_n(\mathbf{t}) = \sum_{i=1}^{n} \mathbf{c}_i \psi(X_i - \mathbf{t}' \mathbf{c}_i), \qquad \mathbf{t} \in E^q. \qquad (5.2.19)$$

For symmetric F, $\mathbf{S}_n(\boldsymbol{\beta})$ has location $\mathbf{0}$, so the estimator $\check{\boldsymbol{\beta}}_n$ of $\boldsymbol{\beta}$ is defined as the solution of

$$\mathbf{S}_n(\check{\boldsymbol{\beta}}_n) = \mathbf{0}. \qquad (5.2.20)$$

For the study of the invariance principles for the $\check{\boldsymbol{\beta}}_n$, it may be quite convenient to study linearity results on $\{\mathbf{S}_k(\boldsymbol{\beta}+n^{-1/2}\mathbf{t})-\mathbf{S}_k(\boldsymbol{\beta}):k\leq n\}$ and combine these with appropriate invariance principles for the $\mathbf{S}_k(\boldsymbol{\beta})$ (which hold under very general regularity conditions). This approach has been studied in detail in Jurečková and Sen (1981a, d), (1984) and for the location model, the details are also provided in Sen (1981d, §§ 8.3 and 8.4). Invariance principles relating to the variance estimators of these estimators are also studied in Sen (1981d).

In all these developments, in Sen (1981d), the basic approach has been through suitable martingale (or reversed martingale) characterizations, which provides comparatively simpler proofs under the usual regularity conditions.

5.3. Asymptotic theory of nonparametric RST. We shall present here the basic theory appropriate for RST, GST and TST procedures. Among these procedures, perhaps the GST has the simplest structure. For a predetermined number k (≥ 1) and nested positive integers $n_1<n_2<\cdots<n_k$ ($=N$), consider suitable test statistics $T_{n_1},\ldots,T_{n_k}$ based on sample sizes $n_1,\ldots,n_k$, respectively. Consider then a set $\{(a_i,b_i),1\leq i\leq k\}$ of real numbers, such that if $T_{n_1}\notin(a_i,b_i)$, stop experimentation at that time along with the rejection of the null hypothesis; otherwise, proceed on and check if $T_{n_2}\in(a_2,b_2)$ or not. In this way, there are at most k tests (interim analysis). The probability of Type I error is therefore

$$P\{T_{n_1}\notin(a_1,b_1)\,|\,H_0\}$$
$$+P\{T_{n_1}\in(a_1,b_1),\,T_{n_2}\notin(a_2,b_2)\,|\,H_0\}+\cdots$$
$$+P\{T_{n_j}\in(a_j,b_j),\,1\leq j\leq k-1,\,T_{n_k}\notin(a_k,b_k)\,|\,H_0\},\qquad(5.3.1)$$

and the basic problem is to determine the moving boundaries (a_i,b_i), $i=1,\ldots,k$, such that (5.3.1) is equal to some pre-assigned α ($0<\alpha<1$). In the case of one-sided tests (say, right-handed), the a_i may be taken to be $-\infty$ ($1\leq i\leq k$).

Typically, in a nonparametric setup, for large sample sizes, the joint distribution of $(T_{n_1},\ldots,T_{n_k})$ (when suitably normalized) may be adequately approximated by a multinormal one, and hence, the problem reduces to that of finding the multinormal probability contents of various rectangular distributions. For $k=2$ or 3, some tables for these are available (viz. Gupta (1963)), while for $k\geq 4$, for specific covariance structures (viz. $n_j=n_j$, $1\leq j\leq k$) and specific sets of (a_i,b_i), some tabulations are due to O'Brien and Fleming (1979). These computations become quite cumbrous as k increases, and for $k\geq 15$, good approximations can be obtained more readily by using Brownian motion approximations. As an illustration, consider a GST symmetry based on the signed rank statistics in (5.2.12). Let then $T_{n_j}=n_j^{-1/2}S_{n_j}/A$, where A^2 is defined before (5.2.13), and let $n_j=jn$, $1\leq j\leq k$, $N=nk$. Then, under H_0, the T_{n_j} have means 0, unit variances and $E(T_{n_j},T_{n_{j'}})=(n_j\wedge n_{j'})/\sqrt{n_jn_{j'}}=(j\wedge j')/\sqrt{jj'}$, for $j,j'=1,\ldots,k$. The O'Brien–Fleming formulae may be adopted in this case. If, however, $k\geq 15$, one may use the weak invariance principles for $\{T_n,n\leq N\}$

(whereby, adopt the Brownian motion approximation) and obtain appropriate critical values by using the tables due to DeLong (1980), (1981), and others. These formulae become quite simple if one chooses $a_j = (j/k)^{-1/2}a$ and $b_j = (j/k)^{-1/2}b$, for $j = 1, \dots, k$. A similar case holds with the other nonparametric statistics considered in § 5.2. For those statistics which are not genuinely distribution-free, the standardized forms (T_{n_j}) would require the estimates of the variance, and hence, we need to assume that these estimates are consistent too.

We consider next the theory of TST procedures. Unlike the case of GST, here, we need to consider the complete set $\{T_n : n \leq N\}$ and formulate a sequential procedure for which the stopping time is bounded from above by N, with probability 1. Typically, in an asymptotic setup, one may be able to reduce the problem to that of a Wiener process as follows. By virtue of (5.2.10) (and parallel results for other nonparametric statistics), consider a standard Brownian motion process on $[0, 1]$. Under the null hypothesis, we have a null drift function, while under a local alternative (relative to the target sample size, N), we have an appropriate drift function (usually linear in this context). Thus, one can adapt the theory of TST for a drifted Wiener process and find out the desired procedure: The boundaries obtained in this manner are generally linear, though some curvilinear ones may also be advocated. With these boundaries, one may switch back to the original sequence $\{T_n, n \leq N\}$ and carry out the TST in a convenient way. Again, as an illustration, consider the test for symmetry based on the signed-rank statistics $\{S_n\}$ in (5.2.12). Here, we take (for a given N), $T_n = N^{-1/2}S_n/A$, $n \leq N$, so that under H_0, $\{T_n, n \leq N\}$ can be approximated, in distribution, by a standard Brownian motion $W = \{W(t), 0 \leq t \leq 1\}$. Under a local alternative (viz. $H_n : F$ is symmetric around

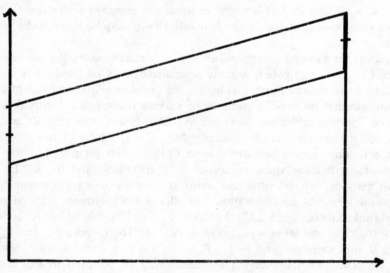

FIG. 5.3.1.

$\theta_N = N^{-1/2}\lambda$, for a given λ), we have a parallel weak convergence result involving a linear drift function $\xi = \{\xi(t) = t\xi, 0 \le t \le 1\}$, where $\xi = \lambda\gamma(F)/A$ and $\gamma(F) = \int_{-\infty}^{\infty} (d/dx)\phi(F(x)) \, dF(x) \, (>0)$.

For a positive ξ, we have two parallel lines with a positive slope, so that whenever $N(t)$ crosses the upper line before crossing the lower one, we stop at that point along with the rejection of H_0; if $W(t)$ goes below the lower line first before crossing the upper one, we stop at that point along with the acceptance of H_0. This is adapted for $t < 1$. At $t = \tau$, we accept or reject H_0 (if no earlier decision were made) according as $W(t)$ is smaller than or greater than the midpoint ξ^*. These two boundaries demand the knowledge of ξ (i.e., λ as well as $\gamma(F)$): one may use sequential estimates of $\gamma(F)$, while λ has to be specified. For a two-sided alternative, we have the picture in Fig. 5.3.2.

Here also, at $t = 1$, we eliminate the continuation region and extend the acceptance region to (ξ_2^*, ξ_1^*) and the complementary set is taken as the rejection region. Noting that T_n corresponds to $W(n/N)$, $n \le N$, the pictures tell us how to use the TST based on the T_n.

The knowledge of λ is necessary in the proper demarcation of the acceptance, rejection and continuation spaces leading to given (asymptotic) levels of Type I and Type II errors. In a practical problem, often, the null hypothesis is well defined, though the alternative ones may not be so (e.g., $\lambda > 0$ but not precisely known). In such a case, instead of controlling both the Type I and Type II errors, one may control only the Type I error and seek to make the

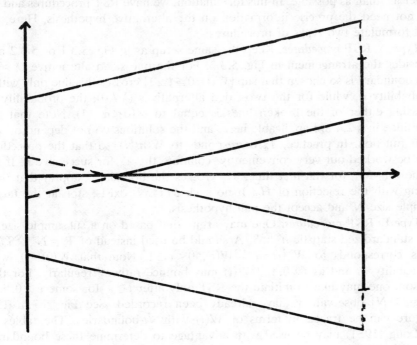

FIG. 5.3.2.

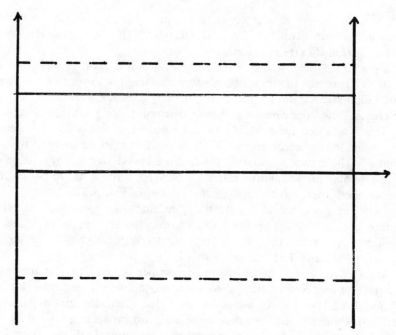

FIG. 5.3.3.

other as small as possible. In this formulation, we have RST procedures and we do not need the precise information on the alternative hypothesis. Here, we will formulate two types of procedures:

Type A RST *procedures*. Keep the same setup as in Fig. 5.3.1 or 5.3.2 and consider the arrangement in Fig. 5.3.3. For the one-sided alternative $(\lambda > 0)$, the boundary is so chosen that $\sup \{W(t): 0 \le t \le 1\}$ crosses this line only with a probability α, while for the two-sided alternatives $(\lambda \ne 0)$, the probability of crossing either of the broken lines is equal to α $(0 < \alpha < 1)$. Note that the formulae in § 2.4 are applicable here, and the solutions do not depend on the drift function. In practice, T_n corresponds to $W(n/N)$, so that the procedure can be worked out very conveniently: compute the T_n for successive n. If for some n $(=M)$, for the first time, T_M crosses the boundary, stop at that time along with the rejection of H_0. If no such M $(\le N)$ exists, stop at the target sample size N and accept the null hypothesis.

Type B RST *procedures*. One may argue that based on a subsample size n, the standardized statistic $n^{-1/2} S_n / A$ should be used instead of $T_n = N^{-1/2} S_n / A$. This corresponds to $W^*(t) = t^{-1/2} W(t)$, $0 < t \le 1$. Note that $W(0) = 0$, with probability 1, and as $t \downarrow 0$, $t^{-1/2} W(t)$ may bounce rather irregularly. For this reason, one may like to initiate the RST only after $t \ge \varepsilon$, for some $\varepsilon > 0$, i.e., when $[\varepsilon N]$ observations have already been recorded. See Fig. 5.3.4. This picture can be traced in terms of $W(t)$ with $\sqrt{t}$-boundaries. The tables in DeLong (1981) may be used with advantage to determine these boundaries corresponding to given level of Type I error. Comparing Figs. 5.3.3 and

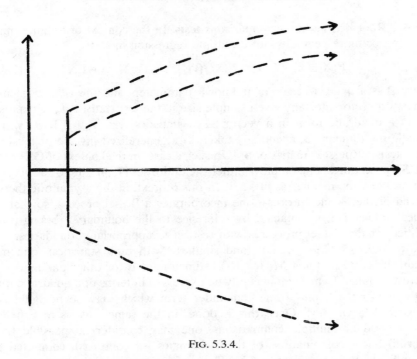

FIG. 5.3.4.

5.3.4, we may remark that whereas for small t (i.e., n/N), the Type B procedure may allow more chance for rejection of H_0 (and early stopping), for t not so small, these square root boundaries are beyond the parallel line ones in Fig. 5.3.3, so that the Type A procedure may have more advantage. Thus, depending on whether we anticipate a significant divergence at the lower or upper range of n ($\leq N$), the Type B or A procedure may be advocated. In this context of RST, generally we have a linear drift (for local alternatives), and hence, the picture may depend well on λ, ε and $\gamma(F)/A$. Numerical and simulation studies have indicated that these two types compare very evenly with each other.

In all the cases depicted in Figs. 5.3.1–5.3.4, instead of the Brownian motion processes, in an asymptotic setup, one may end up with Bessel processes (with appropriate drifts). For such Bessel processes, boundary crossing probabilities (with respect to horizontal as well as square root boundaries) have also been studied in fair detail (cf. DeLong (1980), (1981)), and these may be incorporated in the study of the asymptotic theory of the RST procedures considered. In particular, for local alternatives, the asymptotic power functions as well as the asymptotic distributions of the stopping times can be studied by reference to the first passage problem relating to such drifted Brownian motions or Bessel processes (with respect to horizontal or square root boundaries). There is a good scope for detailed numerical studies for the various specific statistics treated in § 5.2.

5.4. Repeated multiple comparisons tests. In the context of a multisample problem, or more generally, in a multiple regression model:

$$P\{X_i \ge x\} = F_i(x) = F(x - \boldsymbol{\beta}'\mathbf{c}_i), \qquad i \ge 1, \quad x \in E, \qquad (5.4.1)$$

where $\boldsymbol{\beta}$ is a q (≥ 1)-vector of unknown parameters and the $\mathbf{c}_i$ are specified regression vectors, for any given sample size n, the test statistic T_n is expressible as a quadratic form in a vector $\mathbf{S}_n$ of statistics (viz. $T_n = \mathbf{S}_n' \mathbf{B}_n \mathbf{S}_n$ with a suitable discriminant $\mathbf{B}_n$, generally taken as a generalized inverse of the (null hypothesis) dispersion matrix of $\mathbf{S}_n$). In such a case, in the context of GST, TST or RST procedures, one employs either the entire set $\{T_n, n \le N\}$ or a subset of that, in the same manner as in § 5.3. In this context, in the asymptotic theory, instead of the Wiener process, one incorporates a Bessel process, so that the critical values can be obtained by reference to the boundary crossing probabilities for the Bessel processes with respect to appropriate boundaries.

Now as in (5.2.19)–(5.2.20) (and similarly with rank statistics), one may obtain the corresponding M- (or R-) estimators of $\boldsymbol{\beta}$ (or other parameters of interest), so that T_n may equivalently be expressed in terms of a quadratic form in these estimates (plus some remainder term which may be negligible, in probability, as $n \to \infty$). Once this is done, in the same way as in Scheffé's (S-) method of multiple comparisons, one may consider all possible (normalized) linear combinations of these estimates (or contrasts), computed for the individual sample sizes $n \le N$. Thus, with respect to the critical levels, computed under the specified GST, RST or TST rule, one may have a simultaneous testing procedure as well as a multiple comparisons rule. These procedures are natural extensions of the S-procedure; instead of the conventional chi-square or variance-ratio distributions, one needs to consider the corresponding Bessel process distributional results. In the special case of some balanced designs, Tukey's T-method of multiple comparisons (involving the studentized range of some normal variables) can be similarly extended to a GST, TST or RST problem, where we need to consider the distribution of the maximum (over $n \le N$) of the range of a vector of Wiener processes. Some of these results are more useful in the context of multiple paired comparisons procedures, in the context of a GST, TST or RST. We may refer to So and Sen (1982a,b) for the case of rank and M-procedures, for the several samples problems. Similar solutions work out well for other statistics as well. The basic theory is again based on the invariance principles discussed in § 5.2.

5.5. Some general remarks. In this chapter, the main emphasis is laid on the testing aspects in the context of interim analysis. Side by side, the problem of estimating the parameters, following the termination of a sequential or quasi-sequential procedure, remains of genuine interest. Since the stopping rule, in general, is interrelated with the estimation rule, it may be quite difficult to prescribe a simple formula for such "randomly stopped estimation rules". Obviously, the sample size(s) on which the estimators are to be based will no longer be fixed, and the probability law governing the distribution of the sample size(s) depends on that of the stopping rule. For the testing problem, subject to a given margin of Type I error, earlier stopping of the trial is welcome; on the other hand, from the estimation point of view, the smaller the (stopped) sample size, the poorer may be the estimators based on it. This generally introduces more variability in such quasi-sequential estimators. However, as in § 5.4, for various values of n, one may derive the estimators from the corresponding statistics (T_n), so that with respect to the overall critical levels for the set $\{T_n, n \leq N\}$ (computed for a given GST, TST or RST rule), one may get a confidence set for the parameters under consideration. This procedure allows for a possible early termination and has an (asymptotic) lower bound for the coverage probability. Further, as in § 5.4, one may use this procedure for attaching a (quasi-sequential) simultaneous confidence set to the parameters under consideration. We may again refer to So and Sen (1982a, b) for some nonparametric procedures for this problem.

Basically, in a GST, TST or RST procedure, if we use a sequence $\{T_n\}$ of genuinely distribution-free statistics (e.g., the sequence of signed-rank statistics for the hypothesis of symmetry) then the procedure remains distribution-free as well. This characterization may not be applicable to the procedures based on linear combinations of functions of order statistics, M-statistics or U-statistics (in a general setup). In such a case, one needs to estimate also the (null) variance function of the associate sequence of statistics, and the consistency of these estimators (for all $n \leq N$) may require to wait until at least a few observations are available and then to initiate the GST/TST/RST procedures. In a practical setup, this does not, however, pose any severe restriction. If N is comparatively large, one would not like to make a decision until at least a few observations are available. On the other hand, if there is a significant departure from the null hypothesis, in a RST, one would expect an early termination with a high probability, and hence, the initial sample size (n_0) to resume RST should not be very large. It is difficult to prescribe a general formula for choosing such an n_0 (in relation to N). However, some simulation studies (made in some specific cases) suggest that for a Type B RST procedure, n_0 may be taken as the minimum sample size needed to justify the normality results on T_{n_0}. For a Type A RST procedure, n_0 does not play a vital role, and may even be taken smaller. In the context of estimation following a RST, one of course may even choose a larger n_0.

The asymptotic theory discussed in this chapter is primarily based on the invariance principles for the various nonparametric statistics (described in § 5.2) and stronger modes of convergence of their variance estimators. However, in this context, we have not studied any suitable rates of convergence for these asymptotic results, and hence, we are not in a position to prescribe a formula for the adequacy of these asymptotic approximations for moderately large sample sizes N. Actually, this will generally depend on a number of factors, viz. the sequence $\{T_n\}$, the d.f. F (unless we have a genuinely distribution-free procedure) and the variance estimators as well. Numerical studies have indicated that in some of the distribution-free cases, the asymptotic theory leads to a conservative procedure (i.e., the actual Type I error is smaller than the prescribed value); the situation with the M-procedures may be somewhat different, but their robustness picture remains intact. In this respect, more theoretical studies are needed and these are contemplated for future works.

We conclude this chapter with the remark that in a RST/TST/GST, the stopping number $(K/K^*/Km)$ is bounded from above, by the target sample size N. As such, for $N^{-1}K$ converging in law to a nondegenerate r.v. τ, we have also the convergence of $E(N^{-1}K)$ to $E(\tau)$. Thus, the expected stopping times (under null as well as local alternatives) may be studied under the same regularity conditions. As such, the Bessel process approximations in § 5.3 also provide information on the asymptotic properties of stopping times for the RST and other procedures.

CHAPTER 6

Nonparametrics of Sequential Tests

6.1. Introduction. For testing a null hypothesis H_0 against an alternative H_1, one may desire to have a test with the *strength* (α, β), for some preassigned $0 < \alpha < 1$ and $0 < \beta < 1$, that is, a test with the probabilities of Type I and Type II errors bounded by α and β, respectively. When both H_0 and H_1 are simple hypotheses, it is usually possible to achieve this by choosing the sample size adequately large, and using a fixed sample size test. However, when H_0 and H_1 are not simple hypotheses (so that the probability law depends on some nuisance parameter(s)), such a fixed sample size test with the prescribed strength may not exist. For the normal mean testing problem, Stein (1945) considered a two-stage (sequential) procedure whose power is independent of the nuisance parameter (σ). Even for the case of simple H_0 vs. simple H_1, though we may have a fixed sample size test of prescribed strength (α, β) (when the sample size is chosen adequately large), this test may not be the fully efficient one (in the sense that there may be alternative tests with the same strength which may require, on an average, a smaller number of observations). Indeed, Wald's (1947) *sequential probability ratio tests* (SPRT), besides having the prescribed strength (α, β), require, on an average, a smaller number of observations than the corresponding Neyman–Pearson theory based on optimal fixed sample size procedures. The optimality of SPRT, established by Wald and Wolfowitz (1948) bears this testimony in a more general setup.

When H_0 and H_1 are not simple hypotheses, the situation becomes a bit more complicated due to the presence of nuisance parameters. Among other generalizations of SPRT, the *sequential likelihood ratio tests* (SLRT), due to Bartlett (1946) and Cox (1963), among others, though of asymptotic character, possess remarkable adaptability in a variety of situations. These are conceptually the precursors of nonparametric sequential tests. The (asymptotic) theory of SLRT rests on some (stronger) invariance principles for the likelihood ratio statistics. Similar invariance principles hold for a veriety of nonparametric statistics and these enable us to develop the theory on nonparametric sequential tests in the same level of generality as of SLRT's. These invariance principles (for nonparametric statistics) have been studied in detail in Sen (1981d, Part I) and incorporated in Chapter 9 in the study of the asymptotic theory of nonparametric sequential tests and the related *asymptotic relative efficiency* (ARE) results. As in Sen (1981d, Chap. 9), we shall find it convenient to present the theory of nonparametric sequential tests based on suitable estimates and aligned statistics, and to unify them in the light of similarity of their OC and ASN functions.

Borrowing the basic idea from the normal theory SLRT, in § 6.2, we consider estimators based sequential tests, and discuss the basic nonparametric feature of these procedures. Basic results on the termination (with probability 1 of these sequential tests and their asymptotic OC and ASN functions are also discussed. Parallel results for the sequential tests based on aligned non-parametric statistics are considered in § 6.3. In this context, it may be remarked that for either type of these sequential tests, to study the asymptotic form of the ASN (requiring the first order moment convergence of stopping times) one may need regularity conditions comparatively more stringent than the ones needed to study the asymptotic OC function (requring only the convergence in probability results). However, to study the ARE of two competing sequential tests, for the common testing problem, both having the same asymptotic OC function, one may essentially compare the asymptotic distributions of the two *stopping times*; this provides a natural generalization of the classical (nonsequential) Pitman-ARE to the sequential case, and for this purpose, we do not need any extra regularity condition. Following Sen and Ghosh (1980), we present a general account of this sequential Pitman-ARE in § 6.4. An account of nonparametric (sequential) tests with power 1 (along with the relevant invariance principles) is given in § 6.5. The concluding section deals with some general remarks and ties up the connection with the other tests in Chapter 5.

6.2. Nonparametric estimators based sequential tests. To motivate the theory, let us consider the SPRT for the normal mean testing problem (variance known). Let $\{X_i; i \geq 1\}$ be a sequence of i.i.d.r.v.'s having the normal d.f. with mean μ (unknown) and a known variance $\sigma^2 (<\infty)$. Suppose that we want to test for $H_0: \mu = \mu_0$ vs. $H_1: \mu = \mu_1 = \mu_0 + \Delta$, where μ_0 and Δ are specified. In the Wald (1947) SPRT setup, we conceive of two constants $b = \log B$ and $a = \log A$ (where $0 < B < 1 < A < \infty$), and for every $n \geq 1$, define the log-probability ratio as

$$\log \lambda_n = \Delta \sigma^{-2} \left(\sum_{i=1}^{n} \{X_i - \tfrac{1}{2}(\mu_0 + \mu_1)\} \right). \tag{6.2.1}$$

Then, one continues drawing observations one by one as long as

$$b < \log \lambda_n < a, \qquad n \geq 1. \tag{6.2.2}$$

If, for the first time, for $n = N$, $\log \lambda_n \notin (b, a)$, sampling is stopped at that stage, and H_0 or H_1 is accepted according as $\log \lambda_n$ is $\leq b$ or $\geq a$. If no such N exists, we let $N = +\infty$. The *stopping variable* N is therefore a positive integer valued r.v. The constants A and B are to be so chosen that

$$P\{\text{Type I error}\} = P\{H_0 \text{ rejected} \mid H_0\} \leq \alpha, \tag{6.2.3}$$

$$P\{\text{Type II error}\} = P\{H_0 \text{ accepted} \mid H_1\} \leq \beta, \tag{6.2.4}$$

where $(0<) \alpha, \beta (<1)$ are preassigned numbers. Actually,

$$B \geq \beta/(1-\alpha) \quad \text{and} \quad A \leq (1-\beta)/\alpha \tag{6.2.5}$$

and for small Δ, the inequality signs may as well be replaced by $=$ signs.

Consider next the Bartlett–Cox SLRT for the same testing problem (arising in the more likely case where σ^2 is unknown). For every $n \geq 2$, define $s_n^2 = (n-1)^{-1} \sum_{i=1}^{n} (X_i - \bar{X}_n)^2$, $\bar{X}_n = n^{-1} \sum_{i=1}^{n} X_i$, and in (6.2.1), consider the modified definition

$$\log \lambda_n = \Delta s_n^{-2} \left(\sum_{i=1}^{n} \{X_i - \tfrac{1}{2}(\mu_0 + \mu_1)\} \right), \qquad n \geq 2. \tag{6.2.6}$$

Intuitively, s_n^2 a.s. converges to σ^2 (as $n \to \infty$), while (even if F is not a normal d.f. but admits of a finite second order moment) the Skorokhod–Strassen embedding of a Wiener process remains adaptable for $\{\sum_{i=1}^{n} (X_i - \mu)/\sigma\}$. Thus, whenever Δ is small, one may conceive of a drifted Wiener process with a linear drift, incorporate the sequential testing theory of Dvoretzky, Kiefer and Wolfowitz (1953) and formulate the same procedure for the sequence in (6.2.6). This feature is quite general to yield asymptotic approximation to the OC function, though for the ASN (requiring the computation of EN), one may need additional regularity conditions to ensure the desired moment-convergence.

For a general class of estimators, the Skorokhod–Strassen embedding of a Wiener process holds under quite general regularity conditions. Also, a.s. convergence of their (asymptotic) variance estimators holds under similar conditions. These enable us to formulate a general class of sequential tests without requiring the form of the underlying d.f. F to be specified. We shall consider this class here. Based on $\mathbf{X}^{(n)} = (X_1, \ldots, X_n)$, let $T_n = T(\mathbf{X}^{(n)})$ be an estimator of a real valued parameter θ, defined for every $n \geq n_0$, for some $n_0 (\geq 2)$. Consider the hypotheses:

$$H_0 : \theta = \theta_0 \quad \text{vs.} \quad H_1 : \theta = \theta_1 = \theta_0 + \Delta \tag{6.2.7}$$

where θ_0 and Δ are specified. Suppose further that

$$\sqrt{n}(T_n - \theta)/\sigma_\theta \sim \mathcal{N}(0, 1) \quad \text{as } n \to \infty, \tag{6.2.8}$$

where σ_θ $(0 < \sigma_\theta < \infty)$ is a continuous function of θ in some neighborhood of θ_0. Generally, σ_θ^2 is unknown, and we assume that there exists a sequence $\{s_n^2\}$ of (strongly) consistent estimators of σ_θ^2, where $s_n^2 = s^2(\mathbf{X}^{(n)})$ is based on $\mathbf{X}^{(n)}$, for $n \geq n_0$. By analogy with (6.2.6), we now define

$$Z_n = n\Delta(T_n - \tfrac{1}{2}(\theta_0 + \theta_1))/s_n^2, \qquad n \geq n_0. \tag{6.2.9}$$

Then, as in (6.2.2), we continue sampling (starting with the initial sample size n_0) as long as Z_n in (6.2.9) $\in (b, a)$, $n \geq n_0$. If, for the first time, for $n = N$ $(\geq n_0)$, $Z_n \notin (b, a)$, we stop sampling, and we accept H_0 or H_1 according as $Z_n \leq b$ or $\geq a$. Thus, N is the stopping number. Note that, by definition, for every $n \geq n_0$,

$$P\{N > n\} = P\{Z_m \in (b, a), \forall \, m : n_0 \leq m \leq n\} \leq P\{Z_n \in (b, a)\}, \tag{6.2.10}$$

so that on noting that for every (fixed) Δ, $n^{-1/2} s_n^{-1} a$ and $n^{-1/2} s_n^{-1} b$ both (a.s.) converge to 0 as $n \to \infty$, while by (6.2.8) and the stochastic convergence of s_n to

σ_θ (as has been assumed), the right-hand side of (6.2.10) converges to 0, as $n \to \infty$. Thus, the *procedure terminates with probability* 1.

Note that for every (fixed) $\theta \neq \frac{1}{2}(\theta_0 + \theta_1)$ $(= \theta_0 + \frac{1}{2}\Delta, \Delta$ fixed$)$, $|n\Delta(\theta - \frac{1}{2}(\theta_0 + \theta_1))| \to \infty$ as $n \to \infty$, and hence, it can be verified that the OC function (i.e, the probability of accepting H_0 when θ is the true parameter value) converges to 1 or 0, according as θ is $<$ or $> \frac{1}{2}(\theta_0 + \theta_1)$. To avoid this degeneracy of the OC function, we take recourse to an asymptotic setup where Δ is made to converge to 0. In practice, this setup works out well when Δ is small. We assume that as $\Delta \to 0$,

$$\theta = \theta_0 + \phi\Delta, \qquad \phi \in \Phi = \{\phi : |\phi| \leq K\} \tag{6.2.11}$$

for some K $(< \infty)$. Basically, this amounts to choosing local alternatives. Secondly, we may allow the initial sample size n_0 $(= n_0(\Delta))$ to depend on Δ, in such a way that as $\Delta \to 0$,

$$\Delta^2 n_0(\Delta) \to 0 \quad \text{but} \quad n_0(\Delta) \to \infty. \tag{6.2.12}$$

Regarding the sequence $\{s_n^2\}$, we assume that

$$s_n^2 \to \sigma_\theta^2 \quad \text{a.s., as } n \to \infty, \qquad \lim_{\theta \to \theta_0} \sigma_\theta^2 = \sigma_{\theta_0}^2 < \infty. \tag{6.2.13}$$

Further, for every Δ (> 0), we consider a right-continuous, nondecreasing and integer valued function $k_\Delta(t) = \max\{k : k \leq \Delta^{-2}t\}$, $t \geq 0$, and define a stochastic process $W_\Delta = \{W_\Delta(t), t \geq 0\}$, by letting

$$W_\Delta(t) = \Delta k_\Delta(t)\{T_{k_\Delta(t)} - \theta\}/\sigma_{\theta_0}, \qquad t \geq 0, \tag{6.2.14}$$

and assume that for every finite R $(< \infty)$, as $\Delta \downarrow 0$,

$$W_\Delta^T = \{W_\Delta(t); 0 \leq t \leq T\} \underset{\mathscr{D}}{\to} \{W(t); 0 \leq t \leq T\} \tag{6.2.15}$$

where $\{W(t), t \geq 0\}$ is a standard Wiener process.

Note that we may rewrite Z_n in (6.2.9) [under (6.2.10)] as

$$\begin{aligned} Z_n &= n\Delta(T_n - \theta)/s_n^2 + n\Delta(\theta - \frac{1}{2}(\theta_0 + \theta_1))/s_n^2 \\ &= (\sigma_\theta^2/s_n^2)\{n\Delta(T_n - \theta)/\sigma_\theta^2 + n\Delta^2(\phi - \frac{1}{2})/\sigma_\theta^2\}, \end{aligned} \tag{6.2.16}$$

so that for $n = k_\Delta(t)$, as $\Delta \downarrow 0$, by (6.2.13)–(6.2.15), we have the distributional approximation for $Z_{k_\Delta(t)}$ as

$$\sigma_{\theta_0}^{-1} W(t) + (\phi - \frac{1}{2})t\sigma_{\theta_0}^{-2} = \sigma_{\theta_0}^{-1}\{W(t) + \sigma_{\theta_0}^{-1}t(\phi - \frac{1}{2})\}, \qquad t \geq 0. \tag{6.2.17}$$

Consequently, we may approximate the OC function by that of the probability that a Wiener process with the drift function $\{t(\phi - \frac{1}{2})/\sigma_{\theta_0}, t \geq 0\}$ will cross the line $b\sigma_{\theta_0}$ before crossing the upper boundary $a\sigma_{\theta_0}$, and for this the standard results in Dvoretzky, Kiefer and Wolfowitz (1953) apply, and lead us to the following:

Under (6.2.11) through (6.2.15), the limiting OC function $(= P(\phi))$ of the sequential test is given by

$$P(\phi) = \begin{cases} (A^{1-2\phi} - 1)/A^{1-2\phi} - B^{1-2\phi}), & \phi \neq \frac{1}{2}, \\ a/(a - b), & \phi = \frac{1}{2}, \end{cases} \tag{6.2.18}$$

when A and B are defined by (6.2.5) with the equality sign, and $a = \log A$, $b = \log B$. Thus, $P(0) = (A - 1)/(A - B) = 1 - \alpha$ and $P(1) = \beta$, i.e., the asymptotic strength of the test is (α, β).

For the study of the limiting ASN function, our goal is to show that

$$\lim_{\Delta \downarrow 0} \Delta^2 EN(\Delta) = \begin{cases} (bP(\phi) + a\{1 - P(\phi)\})\sigma_{\theta_0}^2/(\phi - \tfrac{1}{2}), & \phi \neq \tfrac{1}{2}, \\ -ab\sigma_{\theta_0}^2, & \phi = \tfrac{1}{2}, \end{cases} \quad (6.2.19)$$

where $N(\Delta) = \inf\{n \geq n_0 : Z_n \notin (b, a)\}$ and $P(\phi)$ is defined by (6.2.18). The computation of $EN(\Delta)$ and its convergence, naturally, may demand stronger regularity conditions. First, we replace the first part of (6.2.13) by the more stringent condition that there exists a $\delta > 0$, such that for every $\eta > 0$ and n adequately large,

$$P_\theta\{|s_n^2/\sigma_\theta^2 - 1| > \eta\} \leq kn^{-1-\delta}, \quad \text{for some } K < \infty. \quad (6.2.20)$$

Further, we assume that for every θ belonging to some neighborhood of θ_0, there exist a sequence $\{T_{n,\theta}^*\}$ and a sequence $\{e_n\}$, such that

$$P_\theta\{|n(T_n - \theta) - T_{n,\theta}^*| > e_n\} \leq Kn^{-1-\delta}, \quad (6.2.21)$$

where $\{T_{n,\theta}^*\}$ is a martingale, $E_\theta T_{n,\theta}^* = 0$, $\forall n \geq 1$,

$$n^{-1}E(R_{n,\theta}^*)^2 \to \sigma_\theta^2 \quad \text{as } n \to \infty, \quad (6.2.22)$$

$$n^{-1}E\left\{\max_{n_0 \leq k \leq n} |T_{k,\theta}^* - T_{k-1,\theta}^*|^2\right\} \to 0 \quad \text{as } n \to \infty, \quad (6.2.23)$$

and, for some $r > 2$,

$$\overline{\lim}\{n^{-r/2}E_\theta|T_{n,\theta}^*|^r\} < \infty; \quad (6.2.24)$$

while $n^{-1/2}e_n \to 0$ as $n \to \infty$. Under these additional conditions, we may virtually repeat the proof of Sen (1981d, Thm. 9.6.1) and conclude that (6.2.19) holds. We may remark, however, that (6.2.19) is not needed for the study of the ARE, and this will be elaborately studied in § 6.4.

We illustrate the basic results of this section with some concrete examples. First, consider the case of general estimable parameters for which Hoeffding's (1948) U-statistics are optimal estimators. For U-statistics, we may use the jackknifed variance estimator (see (3.2.12), for example). Invariance principles for U-statistics (viz. Miller and Sen (1972) and Sen (1974b)) ensure (6.2.15), while (6.2.13) follows from the a.s. convergence properties of jackknifed variance estimators. Thus, (6.2.18) holds under the usual regularity conditions (pertaining to (6.2.8)). Hoeffding's (1961) decomposition leads to an easy verification of (6.2.21), while (6.2.20) follows under the assumption that the kernel has finite moments up to the rth order, for some $r > 4$. (6.2.22)–(6.2.24) are also easily verifiable using the martingale structure of the Hoeffding decomposition (see Sen (1973a)). Thus, (6.2.15) holds. In particular, the SLRT in (6.2.6) corresponds to the particular case, where the degree of the kernel is equal to 1, so that the limiting OC (and ASN) results remain valid for all F for which X has finite second (and rth order, $r > 4$) moments.

As a second example, we may consider the case of sequential tests based on L-estimators. If $X_{n:1} < \cdots < X_{n:n}$ is the ordered r.v. of a sample of size n from a d.f. F, then typically an L-statistic is of the form

$$L_n = \sum_{i=1}^{n} c_{ni} X_{n:i}, \qquad n \geq 1 \tag{6.2.25}$$

where the c_{ni} are given constants. (For an L-estimator of location, the c_{ni} are all nonnegative and $\sum_{i=1}^{n} c_{ni} = 1$.) Using some reverse martingale approximations, invariance principles for $\{L_n\}$ have been developed in Sen (1978a), and a.s. convergence properties of the related variance estimators have also been studied there; a more unified approach has been laid down in Sen (1981d, Chap. 7). A more recent paper (Sen (1984c)) examines the jackknife estimator of the variance and provides parallel results. All these ensure that (6.2.13) and (6.2.15) hold. Thus, (6.2.18) also holds. Verification of (6.2.20)–(6.2.24) may easily be made by using the stronger results of Sen (1977c), Gardiner and Sen (1979) and Sen (1984c), requiring additional regularity conditions; these ensure (6.2.19). In the above discussion, it has been tacitly assumed that the "weights" c_{ni} lead to some "smooth" score function, for which the refined treatment holds. On the other hand, in practice, one may also use a selected number of order statistics (sample quantiles), so that in (6.2.25), we may not have a "smooth" weight function. In such a case, the desired invariance principles follow more directly by an appeal to the Bahadur (1966) representation of sample quantiles. These results are discussed in detail in Sen (1981d, § 7.3).

Jurečková and Sen (1981b) have studied sequential tests based on M-estimators of location and regression parameters (when the score function may possess finitely many discontinuities); in this context, some of the needed invariance principles (parallel to (6.2.13), (6.2.15), (6.2.20), (6.2.21)–(6.2.24)) were developed in their earlier (1981a) paper.

In the context of the (statistical) strength of a bundle of parallel filaments, Sen (1973b) considered sequential tests based on similar invariance principles. In a problem of reliability, a sequential test based on similar invariance principles (for the estimator of "availability") has been considered by Bhattacharjee and Sen (1984). Basically, in all these situations, once one has the desired invariance principles, the asymptotic theory of sequential tests follows on the same line as in (6.2.18) and (6.2.19). For a more elaborate discussion, we may refer to Sen (1981d, §§ 9.4, 9.5, and 9.6). The theory is of nonparametric nature and is adaptable in a broad class of problems.

6.3. Sequential tests based on aligned statistics. In a simple linear model, the least squares estimator and the related test statistic are related by linear or quadratic functional relationships. Similar relations hold, in an asymptotic setup, for some nonparametric statistics as well. For example, for the simple regression model, the Jurečková (1969) linearity of a linear rank statistic in the regression parameter enables one to derive an asymptotically linear relationship between the rank statistic and the derived R-estimator of regression. A very similar situation arises in the one-sample location model (viz. Sen and Ghosh (1971)). Since, computationally, the rank statistics are generally simpler than the derived R-estimators, and moreover, under suitable hypotheses, these rank statistics are distribution-free, one may be interested in considering a variant form of the statistics in (6.2.9), where in the numerator, instead of the estimator T_n (aligned by $\frac{1}{2}(\theta_0 + \theta_1)$), one may use an aligned statistic (with necessary adjustments for the denominator). This was worked out by Sen and Ghosh (1974) for the location problem, and can be more formally presented as follows:

Suppose that we have the same hypothesis testing problem as in (6.2.7). Let $\{U_n = U(\mathbf{X}^{(n)}), n \geq n_0\}$ be a sequence of real-valued statistics, such that (i) under H_0, U_n has a known d.f. with location μ_0 (known) and a known variance σ^2/n, (ii) under a local alternative (i.e., Δ small), the asymptotic mean of U_n is $\Delta\gamma + o(\Delta)$ for some (possibly unknown) γ ($\neq 0$), a functional of the d.f. F, (iii) for every $n \geq n_0$, and θ, there exists a transformation $\mathbf{X}^{(n)} \to \mathbf{X}_\theta^{(n)}$, $U_n \to U_n(\theta) = U(\mathbf{X}_\theta^{(n)})$, such that when θ holds, $U_n(\theta)$ is distribution-free (i.e., has the same d.f. as of U_n under H_0), and (iv) there exists a sequence $\{D_n\}$ of consistent estimates of γ. We define then $\bar{\theta} = \frac{1}{2}(\theta_0 + \theta_1)$ and let $\bar{U}_n = U_n(\bar{\theta}) = U(\mathbf{X}_{\bar{\theta}}^{(n)})$. Thus, $\bar{U}_n$ is essentially an aligned statistic. With this notation, in (6.2.9), we may consider the alternative form

$$Z_n = \Delta n D_n \bar{U}_n / \sigma^2, \qquad n \geq n_0; \tag{6.3.1}$$

of course, if γ were known, then we would have used γ instead of D_n. With this alteration of (6.2.9), the sequential procedure and the stopping rule are the same as in after (6.2.9). By virtue of (6.2.10), to show that the process terminates with probability 1, here, we need to show that $D_n \xrightarrow{P} \gamma$ as $n \to \infty$ and $\bar{U}_n$ possesses some convergence properties too; these are usually quite easy to verify in specific cases.

For the study of the OC function, again, we take recourse to the asymptotic setup in (6.2.11)–(6.2.12). Since, in (6.3.1), σ^2 is known, while D_n estimates γ, here, instead of (6.2.13), we assume that

$$D_n \to \gamma \quad \text{a.s., as } n \to \infty; \tag{6.3.2}$$

it is, of course, possible to relax (6.3.2) a little: For every $\varepsilon > 0$ and $T < \infty$, as $\Delta \downarrow 0$,

$$\max\{|D_n - \gamma| : n\Delta^2 \in [\varepsilon, T]\} \xrightarrow{P} 0. \tag{6.3.3}$$

Further, parallel to (6.2.14), we define here $W_\Delta(t) = \Delta k_\Delta(t)\bar{U}_{k_\Delta(t)}/\sigma$, $t \in [0, T]$, and assume that under (6.2.11), as $\Delta \downarrow 0$,

$$\{W_\Delta(t), 0 \leq t \leq T\} \xrightarrow[\mathcal{D}]{} \{W(t) + (\phi - \frac{1}{2})t\gamma/\sigma, 0 \leq t \leq T\}, \tag{6.3.4}$$

for every $T < \infty$. Then, we may virtually repeat the steps in (6.2.16)–(6.2.17), with the changes made above, and conclude that under (6.2.9), (6.2.10) and (6.3.3), (6.3.4), the limiting OC function ($= P(\phi)$) of the sequential test is again given by (6.2.18). Thus, the asymptotic strength of the sequential test is again (α, β).

For the study of the ASN in an asymptotic setup, we again verify that (6.2.19) holds with appropriate changes. Towards this parallel to (6.2.13) and (6.2.20), here, we assume that for some $\delta > 0$ and every $\varepsilon > 0$, for all n, adequately large,

$$P\{|D_n/\gamma - 1| > \varepsilon\} \leq Kn^{-1-\delta} \quad \text{for some } K < \infty, \tag{6.3.5}$$

and this, of course, ensures (6.3.3). Parallel to (6.2.21), here we conceive of a sequence $\{h_n(\theta)\}$, such that under (6.2.11),

$$P_\theta\{|n(\bar{U}_n - h_n(\theta)) - U_{n,\phi}^*| > e_n\} \leq kn^{-1-\delta}, \tag{6.3.6}$$

where $n^{-1/2}e_n \to 0$ as $n \to \infty$, $\{U_{n,\phi}^*\}$ is a (zero mean) martingale and it satisfies (6.2.2.22)–(6.2.24), and finally, under (6.2.11),

$$\max \{|h_n(\theta) - \Delta(\phi - \tfrac{1}{2})\gamma| : \varepsilon\Delta^{-2} \leq n \leq T\Delta^{-2}\} = o(\Delta) \tag{6.3.7}$$

for every $0 < \varepsilon < 1 < T < \infty$.

With these additional regularity conditions, one may verify that (6.2.19) holds for the sequential procedure here, where we need to replace $\sigma_{\theta_0}^2$ by σ^2/γ^2; see Sen (1981d, Thm. 9.6.2) in this context.

We illustrate this sequential procedure by means of the following examples. Let $\{X_i; i \geq 1\}$ be a sequence of i.i.d.r.v. with a d.f. F_θ where

$$F_\theta(x) = F(x - \theta), \quad x \in E, \qquad F(x) + F(-x) = 1 \quad \forall x. \tag{6.3.8}$$

Thus, θ is the median of the symmetric d.f. F_θ. For this location model, we consider the hypotheses in (6.2.7), and as in Sen and Ghosh (1974), we consider the following sequential rank tests. Without any loss of generality, we may take $\theta_0 = 0$ and $\theta_1 = \Delta > 0$. For every n (≥ 1) and real b, we consider the signed-rank statistic $S_n(b)$ defined as in (4.3.2). Note that under $H_0: \theta = 0$, $S_n(0)$ has a known (symmetric) distribution with location 0 and variance $nA_n^2 = \sum_{i=1}^n (a_n^*(i))^2$, where $A_n^2 \to A^2 = \int_0^1 \phi^2(u)\,du$ ($< \infty$) as $n \to \infty$. Thus, if we put $U_n = n^{-1}S_n$, then under H_0, U_n has mean 0 and a known variance $n^{-1}A_n^2$. For a local alternative, as in (6.2.11), we may proceed as in Sen (1970) and verify that the asymptotic mean of U_n is $\Delta\gamma + o(\Delta)$, where γ is defined after (4.3.4). Further $S_n(b)$ is $\searrow$ in b and under θ, $S_n(\theta)$ has the same distribution as of $S_n(0)$ under H_0. Finally, we may consider the estimator d_n, defined by (4.3.8), and by virtue of (4.3.9), conclude that $d_n \to \gamma$ a.s., as $n \to \infty$. Hence, we may use the sequence $\{Z_n\}$ in (6.3.1) with $Z_n = n\,d_n\bar{U}_n/A_n^2 = \Delta\,d_nS_n(\tfrac{1}{2}\Delta)/A_n^2$, $n \geq n_0$. The weak invariance principle in (6.3.4) follows from the general results in Sen (1981d, Chap. 5); see also § 5.2 of this monograph and Müller-Funk (1979). Thus, the procedure based on these aligned signed-rank statistics terminates with probability 1, and its asymptotic (as $\Delta \downarrow 0$) OC function is given by (6.2.18). Further (6.3.5) follows from (4.3.9), while (6.3.6) and (6.3.7) may be verified with the aid of the asymptotic a.s. linearity results on $S_n(b)$ (in b "close to" 0), studied in detail by Sen and Ghosh (1971), Sen

(1980a,b) and others. Thus, the asymptotic form of the ASN in (6.2.19) also holds here with $\sigma_{\theta_0}^2 = A^2/\gamma^2$. The two-sample location model (i.e., $X_i \sim F$, $Y_i \sim G$, $F(x) = G(x - \theta)$) can be reduced to the one-sample case by working with the differences $X_i - Y_i$, $i \geq 1$, and hence, the above procedure works out nicely.

Next, we consider sequential rank tests for regression, studied by Ghosh and Sen (1977). Consider the model

$$X_i = \lambda + \theta c_i + e_i, \qquad i \geq 1, \tag{6.3.9}$$

where the c_i are known regression constants, not all equal, (λ, θ) is an unknown parameter vector and the e_i are i.i.d.r.v. with an unknown d.f. F. We intend to test for the hypotheses in (6.2.7), treating λ as a nuisance parameter and without assuming the form of F to be specified. Without any loss of generality, we may take $\theta_0 = 0$. For every n (≥ 1), define a linear rank statistic $T_n = T(\mathbf{X}^{(n)})$ as in (5.2.3)–(5.2.4). For every real b, define $X_i(b) = X_i - bc_i$, $i \geq 1$, and let $T_n(b)$ be the linear rank statistic based on $X_i(b)$, $1 \leq i \leq n$. Then $T_n(b)$ is $\searrow$ in b, and, under θ, $T_n(\theta)$ has the same distribution as of $T_n(0)$ under $H_0: \theta = 0$, where $T_n(0)$ is distribution-free (under H_0) with mean 0 and variance $C_n^2 A_n^2$, with $C_n^2 = \sum_{i=1}^n (c_i - \bar{c}_n)^2$ and $A_n^2 = (n-1)^{-1} \sum_{i=1}^n [a_n(i) - \bar{a}_n]^2$, $\bar{a}_n = n^{-1} \sum_{i=1}^n a_n(i)$. Thus, for $U_n = C_n^{-2} T_n$, the conditions (i), (ii) and (iii) stated before (6.3.1) are easy to verify. The estimator D_n of γ is based on the same principle as in (4.3.8), and hence, we may proceed as in (5.3.1) and take in this case:

$$Z_n = \Delta D_n T_n(\tfrac{1}{2}\Delta)/A_n^2, \qquad n \geq n_0 \ (\geq 2). \tag{6.3.10}$$

With this sequence $\{Z_n\}$, we may proceed as in the location problem and verify that (6.2.18) and (6.2.19) hold here too. For details, we may refer to Sen (1981d, Chap. 9).

Sequential M-tests for location/regression, considered by Jurečková and Sen (1981b) are based on the same alignment principle; there we also need to estimate σ_ψ^2, defined by (4.3.11), and for this (4.3.16) and (4.3.19) provide the necessary results. SLRT's are also based on such aligned test statistics, and their asymptotic theory follows along the same line as in this section.

6.4. Pitman efficiency: Sequential case. In §§ 6.2 and 6.3, we have noticed that the asymptotic expression for the ASN (see (6.2.19)) depends on $P(\phi)$, a, b and the scale factor $\sigma_{\theta_0}^2$. Note that $P(\phi)$, a and b do not depend on the particular sequence of statistics employed in (6.2.9) or (6.3.1), while $\sigma_{\theta_0}^2$ is the asymptotic variance appearing in (6.2.8). Thus, comparing the limiting ASN functions in (6.2.19) for two rival sequential tests (both having the same asymptotic OC function in (6.2.18)), we obtain a measure of asymptotic relative efficiency (ARE) which agrees with the classical Pitman-ARE (in the non-sequential case). In this respect, we may comment that for the verification of (6.2.19), we may need some regularity conditions which are generally more stringent than the ones pertaining to the OC function in (6.2.18) (see for example (6.2.20)–(6.2.24) and (6.3.5)–(6.3.7)). Further, in the fixed sample case, for the Pitman-ARE, one may not need such extra regularity conditions. This raises the question of the necessity of using (6.2.19) for the computation of the ARE. Indeed, it has been shown by Sen and Ghosh (1980) (and others) that for the Pitman-ARE in the sequential case, one does not need to verify (6.2.19), and obtain the same result under the usual regularity conditions pertaining to (6.2.18). The following is a somewhat simplified version of the results of Sen and Ghosh (1980).

For the hypotheses testing problem in (6.2.7), consider two rival sequential procedures, both having the same limiting OC function, given by (6.2.18), and for every $\Delta(\neq 0)$, denote the corresponding stopping numbers by $N(\Delta)$ and $N^*(\Delta)$, respectively. Note that under (6.2.11), for every $\phi \in \Phi$,

$$\Delta^2 N(\Delta) \underset{\mathcal{D}}{\to} \tau_\phi \quad \text{and} \quad \Delta^2 N^*(\Delta) \underset{\mathcal{D}}{\to} \tau_\phi^* \qquad (6.4.1)$$

where τ_ϕ and τ_ϕ^* are the *stopping times* for drifted Wiener processes (with drift functions $t(\phi - \frac{1}{2})/\delta$ and $t(\phi - \frac{1}{2})/\delta^*$, $t > 0$, respectively) with respect to the boundaries $(b\delta, a\delta)$ and $(b\delta^*, a\delta^*)$, respectively; a and b are defined as in (6.2.2). Thus, if we define for every $\phi \in \Phi$,

$$P_\phi(t) = P\{\tau_\phi \leq t\} \quad \text{and} \quad P_\phi^*(t) = P\{\tau_\phi^* \leq t\}, \quad t \geq 0, \qquad (6.4.2)$$

then by reference to the distributions of the first exist times for the drifted Wiener processes (viz. Anderson (1960)), we may easily verify that

$$P_\phi^*(t) = P_\phi(ht) \quad \forall \phi \in \Phi, \quad t \geq 0, \qquad (6.4.3)$$

where

$$h = (\delta^*/\delta)^2. \qquad (6.4.4)$$

Thus, h is the classical Pitman-ARE, and by the same concept of equating the two distributions (as in the nonsequential case), we conclude from (6.4.3) and (6.4.4) that h is also the Pitman-ARE in the sequential case. This agrees with the definition of the ARE based on (6.2.19), but, it does not need the extra regularity conditions pertaining to (6.2.19). We may refer to Sen (1981d, § 9.7) for some detailed discussions on this sequential ARE.

6.5. Nonparametric tests with power one. Tests with power 1 have been proposed and studied in a series of papers (for various specific problems) by Darling and Robbins (1967a,b,c), (1968a,b). Basically, these are sequential tests having the two main characteristics that by choosing an appropriately large initial sample size, the Type I error can be made arbitrarily small, and, for every alternative hypothesis (belonging to a suitable family), the test has power 1.

If $\{T_n = T(\mathbf{X}^{(n)}), n \geq 1\}$ denotes the sequence of test statistics, basically, in a nonparametric setup, one may assume that under the null hypothesis H_0, the T_n are distribution-free, and without any loss of generality, we may set their centering constant to be equal to 0. Suppose now that (i) under the true model,

$$T_n \overset{a.s.}{\to} \tau \quad \text{as } n \to \infty \tag{6.5.1}$$

where τ does not vanish when H_0 does not hold, and (ii) there exists a sequence $\{J_n\}$ of intervals containing 0 as an inner point, such that the diameter of $J_n \to 0$ as $n \to \infty$, and under H_0,

$$T_n \in J_n \quad \text{a.s., as } n \to \infty. \tag{6.5.2}$$

Then, one may consider a stopping variable

$$N = \min \{n \geq n_0 : T_n \notin J_n\} = \infty \quad \text{if no such } n \text{ exists,} \tag{6.5.3}$$

where n_0 is an initial sample size. Thus starting with n_0, one may continue drawing observations until for the first time, for some N ($\geq n_0$), $T_N \notin J_N$. If $N < \infty$, we stop at the Nth stage along with the rejection of H_0, while for $N = \infty$, H_0 is not rejected. (6.5.1) and (6.5.2) ensure that the Type I error is small (when n_0 is large) and Type II error is equal to 0 (when $\tau \neq 0$).

Now, for various nonparametric statistics, the a.s. convergence result in (6.5.1) has been studied in detail in Sen (1981d, Part I). Strong invariance principles for various nonparametric statistics have also been studied there in depth—these ensure the law of iterated logarithm, so that (6.5.2) holds, typically, with the diameter of J_n as $O(n^{-1/2}(\log \log n)^{1/2})$. Actually, if the width of the interval J_n be chosen as $O(n^{-1/2}(\log n)^{1/2})$, then sharper probability inequalities for $T_n \notin J_n$ (under H_0) may be incorporated to provide suitable bounds for the Type I error (as a function of the initial sample size n_0). Some of the typical nonparametric tests with power one have been discussed in Sen (1981d, § 9.2), and the general theory presented there may be applicable in a broad class of problems.

6.6. Some general remarks. In the context of the single or two-sample location and Lehmann models, sequential rank order probability ratio tests have been studied by a host of workers. We may refer to Savage and Sethuraman (1972) where other references are cited too. By construction, in such a case, one either assumes that under the alternative hypothesis, the d.f.s are specified, or one takes recourse to a Lehmann (1953) type alternative hypothesis, for which the rank order probabilities do not depend on the underlying d.f.s. In either way, these are rather arbitrary and may not conform to the genuine location model treated in § 6.3. Lai (1975), (1978) has also considered some elegant theorems on sequential rank statistics. In this formulation too, the location model, treated in § 6.3, may not fit, and the approach of Sen and Ghosh (1974) works out better.

The invariance principles for various nonparametric statistics provide the desired asymptotic solutions to the OC and ASN functions. There remains a natural question: How good are these approximations for moderate values of Δ? (Refer to (6.2.7).) Obviously, the asymptotic results may not be approached uniformly in a class of underlying F or uniformly in a class of statistics used.

There is thus a pressing need to study the rate of convergence for these asymptotic expressions, both in terms of analytical methods and numerical studies. For some simple (e.g., location) models and specific d.f.'s, some numerical studies made, so far, are quite encouraging—hopefully, the advent of the modern computer will pave the way for further studies in this direction.

In Chapter 5, we have studied the RST, GST and TST based on appropriate nonparametric statistics. Basically, all of these share the same methodology with the sequential procedures discussed in this chapter. However, in Chapter 5, for the sequential procedures, we have a stopping number, bounded (from above by the target sample size), and hence, weak invariance principles may suffice. On the other hand, for the study of the asymptotic form of the ASN (as in (6.2.19)) of the sequential procedure, we need comparatively stronger regularity conditions (to ensure the first order moment convergence result). With respect to the nonparametric procedures considered in this as well as earlier chapters, we should keep in mind that robustness is the main consideration underlying their general use. For the robustness, interpreted in a local sense (e.g., error contamination around a specified F_0), M-estimators based procedures may have some advantages—see for example, Jurečková and Sen (1982) for some discussions. However, these M-procedures are not generally scale-equivariant. Adaptive M-statistics are likely to play a vital role in this context. On the other hand, for robustness in a global sense, where the d.f. F may belong to a broad class of d.f.'s, it may be safer to use rank procedures. In some cases, these rank procedures are genuinely distribution-free, while the others are only asymptotically distribution-free. There seems to be a good prospect for sequential procedures based on adaptive statistics: Though one may have to sacrifice the exact distribution-freeness for such procedures, the robust-efficiency aspects may dominate the picture, at least in the asymptotic case. The recent results of Jurečková and Sen (1984) and Hušková and Sen (1984) provide good hope for some further developments. Sequential tests for multiparameter hypotheses in a nonparametric setup are also not yet developed to full generality, and more work on this line is anticipated in the near future.

The scope of nonparametric sequential procedures is not confined only to the case of i.i.d.r.v.'s In some stochastic processes too, they may be incorporated under suitable (semi-) martingale-type characterizations. These may provide a lot of fruitful applications in stochastic models arising in biomedical, physical and health sciences. In particular, counting processes are increasingly adopted for drawing inference in such a context, and sequential procedures may fruitfully be adapted to such models.

References

O. O. AALEN (1978), *Nonparametric inference for a family of counting processes*, Ann. Statist., 6, pp. 701–726.

—— (1980), *A model for nonparametric regression analysis of counting processes*, Lecture Notes in Statistics 2, Springer-Verlag, New York, pp. 1–25.

P. I. ANDERSON, Ø. BORGAN, R. D. GILL AND N. KEIDING (1982), *Linear nonparametric tests for comparison of counting processes, with applications (with discussions)*, Internat. Statist. Rev., 50 pp. 219–258.

P. K. ANDERSEN AND R. D. GILL (1982), *Cox regression model for counting processes: A large sample study*, Ann. Statist., 10, pp. 1100–1120.

T. W. ANDERSON (1960), *A modification of the sequential probability ratio test to reduce the sample size*, Ann. Math. Statist., 31, pp. 165–197.

F. J. ANSCOMBE (1952), *Large sample theory of sequential estimation*, Proc. Cambridge Phil. Soc., 48, pp. 600–607.

P. ARMITAGE, (1975), *Sequential Medical Trials*, John Wiley, New York.

P. ARMITAGE, C. K. MCPHERSON C. K. AND B. C. ROWE (1969), *Repeated significance tests on accumulating data*, J. Roy. Statist. Soc., A132, pp. 235–244.

R. R. BAHADUR (1966), *A note on quantiles in large samples*, Ann. Math. Statist., 37, pp. 577–580.

M. S. BARTLETT (1946), *The large sample theory of sequential tests*, Proc. Cambridge Phil. Soc., 42, pp. 239–244.

M. C. BHATTACHARJEE AND P. K. SEN (1984), *Nonparametric estimation of availability under provisions of spare and repair*, Inst. Statist., Univ. North Carolina, Chapel Hill, Mimeo Rep. 1461.

P. K. BHATTACHARYA AND P. J. BROCKWELL (1976), *The minimum of an adaptive process with applications to signal estimation and storage theory*, Z. Wahrsch Verw. Geb., 37, pp. 51–55.

P. K. BHATTACHARYA AND D. FRIERSON (1981), *A nonparametric control chart for detecting small disorders*, Ann. Statist., 9, pp. 544–554.

G. K. BHATTACHARYYA AND R. A. JOHNSON (1968), *Nonparametric tests for shift at unknown time point*, Ann. Math. Statist., 39, pp. 1731–1743.

P. BILLINGSLEY (1968), *Convergence of Probability Measures*, John Wiley, New York.

R. L. BROWN, J. DURBIN AND J. M. EVANS (1975), *Techniques for testing constancy of regression relationship over time*, J. Roy. Statist. Soc., B37, pp. 149–192.

S. K. CHATTERJEE AND P. K. SEN (1973), *Nonparametric testing under progressive censoring*, Calcutta Statist. Assoc. Bull., 22, pp. 13–50.

H. CHERNOFF AND S. ZACKS (1964), *Estimating the current mean of a normal distribution which is subjected to change in time*, Ann. Math. Statist., 35, pp. 999–1018.

Y. S. CHOW AND H. ROBBINS (1965), *On the asymptotic theory of fixed-width sequential interval for the mean*, Ann. Math. Statist., 36, pp. 457–462.

Y. S. CHOW AND K. F. YU (1981), *The performance of a sequential procedure for the estimation of the mean*, Ann. Statist., 9, pp. 184–188.

D. R. COX (1963), *Large sample sequential tests of composite hypotheses*, Sankhyā, Ser. A, 25, pp. 5–12.

—— (1972), *Regression models and life tables (with discussion)*, J. Roy. Statist. Soc., B 34, pp. 187–220.

—— (1975), *Partial likelihood*, Biometrika, 62, pp. 269–276.

G. B. DANTZIG (1940), *On the non-existence of tests of "Students' hypothesis" having power function independent of σ*, Ann. Math. Statist., 11, pp. 186–192.

D. A. DARLING AND H. ROBBINS, (1967a), *Iterated logarithm inequalities*, Proc. Nat. Acad. Sci. USA, 57, pp. 1188–1192.

—— (1967b), *Inequalities for the sequence of sample means*, Proc. Nat. Acad. Sci. USA, 57, pp. 1577–1580.

—— (1967c), *Confidence sequences for mean, variances and median*, Proc. Nat. Acad. Sci. USA, 58, pp. 66–68.

—— (1968a), *Some further remarks on inequalities for sample sums*, Proc. Nat. Acad. Sci. USA, 60, pp. 1175–1182.

—— (1968b), *Some nonparametric sequential tests with power 1*, Proc. Nat. Acad. Sci. USA, 61, pp. 805–809.

D. M. DeLONG (1980), *Some asymptotic properties of a progressively censored nonparametric test for multiple regression*, J. Multivar. Anal., 10, pp. 363–370.

—— (1981), *Crossing probabilities for a square root boundary by a Bessel process*, Comm. Statist. Theor. Math. A, 10, pp. 2197–2213.

—— (1980), *Some asymptotic properties of a progressively censored nonparametric test for multiple regression*, J. Multivar. Anal., 10, pp. 363–370.

A. DVORETZKY, J. KIEFER AND J. WOLFOWITZ (1953), *Sequential decision problems for processes with continuous time parameter. Testing hypotheses*, Ann. Math. Statist., 24, pp. 254–264.

J. C. GARDINER AND P. K. SEN (1979), *Asymptotic normality of a variance estimator of a linear combination of a function of order statistics*, Z. Wahrsch. Verw. Geb., 50, pp. 205–221.

M. GHOSH AND M. MUKHOPADHYAY (1979), *Sequential point estimation of the mean when the distribution is unspecified*, Comm. Statist. Theor. Meth., A8, pp. 637–652.

—— (1980), *Sequential point estimation of the difference of two normal means*, Ann. Statist., 8, pp. 221–225.

—— (1981), *Consistency and asymptotic efficiency of two-stage and sequential estimation procedures*, Sankhyā, Ser. A, 43, pp. 220–227.

M. GHOSH AND P. K. SEN (1971), *Sequential confidence intervals for the regression coefficient based on Kendall's tau*, Calcutta Statist. Assoc. Bull., 20, pp. 23–36.

—— (1972), *On bounded length confidence intervals for the regression coefficient based on a class of rank statistics*, Sankhyā, Ser. A, 34, pp. 33–52.

—— (1973), *On some sequential simultaneous confidence intervals procedures*, Ann. Inst. Statist. Math., 25, pp. 123–134.

—— (1977), *Sequential rank tests for regression*, Sankhyā, Ser. A, 39, pp. 45–62.

—— (1983), *On two-stage James–Stein estimators*, Sequen. Anal., 2, pp. 359–368.

—— (1984), *On asymptotically risk efficient sequential versions of generalized U-statistics*, Sequen. Anal. 3, pp. 233–252.

S. S. GUPTA (1963), *Probability integrals of multivariate normal and multivariate t*, Ann. Math. Statist., 34, pp. 792–828.

J. HÁJEK (1962), *Asymptotically most powerful rank-order tests*, Ann. Math. Statist., 33, pp. 1124–1147.

—— (1968), *Asymptotic normality of simple linear rank statistics under alternatives*, Ann. Math. Statist., 39, pp. 325–346.

J. HÁJEK AND Z. ŠIDÁK (1967), *Theory of Rank Tests*, Academic Press, New York.

W. HOEFFDING (1948), *A class of statistics with asymptotically normal distribution*, Ann. Math. Statist., 19, pp. 293–325.

—— (1961), *The strong law of large numbers for U-statistics*, Inst. Statist., Univ. North Carolina, Chapel Hill, Mimeo Report 302.

—— (1963), *Probability inequalities for sums of bounded random variables*, J. Amer. Statist. Assoc., 58, pp. 13–30.

M. HUŠKOVÁ (1982), *On bounded length sequential confidence intervals for parameter in regression model based on ranks*, Janos Bolyai Math. Soc., Coll. Nonparamet. Infer., 32, pp. 435–463.

M. HUŠKOVÁ AND J. JUREČKOVÁ (1984), *Asymptotic representation of R-estimators of location*, Proc. 4th Pannonian Symp. Math. Statistics, to appear.

——— (1981), *Second order asymptotic relations of M-estimators and R-estimators in two-sample location model*, J. Statist. Plann. Inf., 5, pp. 309–328.

M. HUŠKOVÁ AND P. K. SEN (1984), *On sequentially adaptive asymptotically efficient rank statistics*, Inst. Statist., Univ. North Carolina, Chapel Hill, Mimeo Report 1475.

N. INAGAKI AND P. K. SEN (1985), *On progressively truncated maximum likelihood estimators*, Ann. Inst. Statist. Math, 37 to appear.

S. JOHANSEN (1983), *An extension of Cox's regression model*, Internat. Statist. Rev., 51, pp. 165–174.

J. JUREČKOVÁ (1969), *Asymptotic linearity of a rank statistic in regression parameter*, Ann. Math. Statist., 40, pp. 1889–1900.

J. JUREČKOVÁ AND P. K. SEN (1981a), *Invariance principles for some stochastic processes related to M-estimators and their role in sequential statistical inference*, Sankhyā, Ser. A, 43, pp. 190–210.

——— (1981b), *Sequential procedures based on M-estimators with discontinuous score functions*, J. Statist. Plan. Infer., 5, pp. 253–266.

——— (1982), *M-estimators and L-estimators of location: Uniform integrability and asymptotically risk-efficient-sequential versions*, Sequen. Anal., 1, pp. 27–57.

——— (1984), *On adaptive scale-equivariant M-estimators in linear models*, Statist. Dec. 2, Suppl., pp. 31–46.

J. KIEFER (1959), *K-sample analogue of the Kolmogorov-Smirnov and Cramér-von Mises' tests*, Ann. Math. Statist., 30, pp. 420–447.

R. KOENKER AND G. BASSETT (1978), *Regression quantiles*, Econometrika, 46, pp. 33–50.

J. A. KOZIOL AND D. P. BYAR (1975), *Percentage points of the asymptotic distributions of one and two sample K-S statistics for truncated or censored data*, Technometrics, 17, pp. 507–510.

T. L. LAI (1975), *On Chernoff-Savage statistics and sequential rank tests*, Ann. Statist., 3, pp. 825–845.

——— (1978), *Pitman efficiencies of sequential tests and uniform limit theorems in nonparametric statistics*, Ann. Statist., 6, pp. 1027–1047.

T. L. LAI AND D. SIEGMUND (1977), *A nonlinear renewal theory with applications to sequential analysis*, Ann. Statist., 5, pp. 946–954.

——— (1979), *A nonlinear renewal theory with applications to sequential analysis, II*, Ann. Statist., 7, pp. 60–76.

E. L. LEHMANN, (1953), *The power of rank tests*, Ann. Math. Statist., 24, pp. 23–43.

T. LINDVALL (1973), *Weak convergence of probability measures and random functions in the function space $D[0, \infty)$*, J. Appl. Probab. 10, pp. 109–121.

F. LOMBARD (1981), *An invariance principle for sequential nonparametric test statistics under contiguous alternatives*, South African Statist. J., 15, pp. 129–152.

——— (1983), *Asymptotic distributions of rank statistics in the change point problem*, South African Statist. J., 17, pp. 83–105.

H. MAJUMDAR AND P. K. SEN (1977), *Rank order tests for grouped data under progressive censoring*, Comm. Statist. Theor. Meth. A, 6, pp. 507–524.

——— (1978a), *Nonparametric tests for multiple regression under progressive censoring*, J. Multivar. Anal., 8, pp. 73–95.

——— (1978b), *Nonparametric testing for simple regression under progressive censoring with staggering entry and random withdrawal*, Comm. Statist. Theor. Meth., A, 7, pp. 349–371.

A. T. MARTINSEK (1984), *Sequential determination of estimator as well as sample size*, Ann. Statist, 12, pp. 533–550.

R. G. MILLER, JR. AND P. K. SEN (1972), *Weak convergence of U-statistics and von Mises' differentiable statistical functions*, Ann. Math. Statist., pp. 31–41.

N. MUKHOPADHYAY (1980), *Consistent and asymptotically efficient two-stage procedure to construct fixed-width confidence intervals for the mean*, Metrika, 27, pp. 281–284.

U. MÜLLER-FUNK (1979), *Nonparametric sequential tests for symmetry*, Z. Wahrsch. Verw. Geb., 46, pp. 325–342.

P. C. O'BRIEN AND T. R. FLEMING (1979), *A multiple testing procedure for clinical trials*, Biometrics 35, pp. 549–556.

E. S. PAGE (1954), *Continuous inspection schemes*, Biometrika, 41, pp. 100–115.

—— (1955), *A test for a change in a parameter occurring at an unknown point*, Biometrika, 42, pp. 523–526.

—— (1957), *On problems in which a change of parameter occurs at an unknown time point*, Biometrika, 44, pp. 248–252.

A. N. PETTITT (1979), *A nonparametric approach to the change point problem*, Appl. Statist., 28, pp. 126–135.

H. ROBBINS (1959), *Sequential estimation of the mean of a normal population*, in Probability and Statistics, H. Cramér Vol., Almquist and Wiksell, Uppsala, pp. 235–245.

S. N. ROY (1953), *On a heuristic method of test construction and its use in multivariate analysis*, Ann. Math. Statist, 24, pp. 220–238.

D. RUPPERT AND R. J. CARROLL (1980), *Trimmed least squares estimation in the linear model*, J. Amer. Statist. Assoc., 75, pp. 828–838.

I. R. SAVAGE AND J. SETHURAMAN (1972), *Asymptotic distribution of the log likelihood ratio based on ranks in the two-sample problem*, in Proc. 6th Berkeley Symposium Mathematical Statistics and Probability, L. LeCam et al., eds., pp. 437–458.

E. SCHECHTMAN AND D. A. WOLFE (1981), *Distribution-free tests for the change-point problem*, Tech. Report, Ohio State Univ.

H. M. SCHEY (1977), *The asymptotic distribution of the one-sided Kolmogorov–Smirnov statistic for truncated data*, Comm. Statist. Theor. Meth. A, 6, pp. 1361–1366.

A. SEN AND M. S. SRIVASTAVA (1975), *On tests for detecting changes in mean*, Ann. Statist., 3, pp. 98–108.

P. K. SEN (1959), *On the moments of sample quantiles*, Calcutta Statist. Assoc. Bull., 9, pp. 1–19.

—— (1960), *On some convergence properties of U-statistics*, Calcutta Statist. Assoc. Bull., 10, pp. 1–18.

—— (1969), *On a robustness property of a class of nonparametric tests based on U-statistics*, Calcutta Statist. Assoc. Bull., 18, pp. 51–60.

—— (1970), *On some convergence properties of one-sample rank order statistics*, Ann. Math. Statist., 41, pp. 2140–2143.

—— (1973a). *Asymptotic sequential tests for regular functionals of distribution functions*, Teor. Veroy. Prim., 18, pp. 235–249; Theory Prob. Appl., 18, pp. 226–240.

—— (1973b), *An asymptotically optimal test for the bundle strength of filaments*, J. Appl. Prob., 10, pp. 586–596.

—— (1974a), *Weak convergence of generalized U-statistics*, Ann. Probability, 2, pp. 90–102.

—— (1974b), *Almost sure behavior of U-statistics and von Mises' differentiable statistical functions*, Ann. Statist., 2, pp. 387–395.

—— (1976a), *A two-dimensional functional permutational central limit for linear rank statistics*, Ann. Probability, 4, pp. 13–26.

—— (1976b), *Asymptotically optimal rank order tests for progressive censoring*, Calcutta Statist. Assoc. Bull., 25, pp. 65–78.

—— (1977a), *Some invariance principles relating to jackknifing and their role in sequential analysis*, Ann. Statist., 5, pp. 315–329.

—— (1977b), *Tied-down Wiener process approximations for aligned rank order statistics and some applications*, Ann. Statist., 5, pp. 1107–1123.

—— (1977c), *On Wiener process embedding for linear combinations of order statistics*, Sankhyā, Ser. A, 39, pp. 138–143.

—— (1978a), *An invariance principle for linear combinations of order statistics*, Z. Wahrsch. Verw. Geb., 42, pp. 327–340.

—— (1978b) *Invariance principles for rank statistics revisited*, Sankhyā, Ser. A., 40, pp. 215–236.

—— (1979a), *Weak convergence of some quantile processes arising in progressively censored tests*, Ann. Statist., 7, pp. 414–431.

—— (1979b), *Rank analysis of covariance under progressive censoring*, Sankhyā, Ser. A., 41, pp. 147–169.

—— (1980a), *On almost sure linearity theorems for signed rank order statistics*, Ann. Statist., 8, pp. 313–321.

—— (1980b), *On nonparametric sequential point estimation of location based on general rank order statistics*, Sankhyā, Ser. A, 42, pp. 201–219.

—— (1980c), *Asymptotic theory of some tests for a possible change in the regression slope occurring at an unknown time point*, Z. Wahrsch. Verw. Geb, 52, pp. 203–218.

—— (1981a), *The Cox regression model, invariance principles for some induced quantile processes and some repeated significance tests*, Ann. Statist., 9, pp. 109–121.

—— (1981b), *Asymptotic theory of sometime-sequential tests based on progressively censored quantile processes*, in Statistics and Probability: Essays in Honor of C. R. Rao, G. Kallianpur et al., eds., North-Holland, Amsterdam, pp. 649–660.

—— (1981c), *Rank analysis of covariance under progressive censoring, II*, in Statistics and Related Topics, M. Csörgő et al., eds., North-Holland, Amsterdam, pp. 285–295.

—— (1981d), *Sequential Nonparametrics: Invariance Principles and Statistical Inference*, John Wiley, New York.

—— (1981e), *On invariance principles for LMP conditional test statistics*, Calcutta Statist. Assoc. Bull., 30, pp. 41–56.

—— (1982a), *Invariance principles for recursive residuals*, Ann. Statist., 10, pp. 327–312.

—— (1982b), *Asymptotic theory of some tests for constancy of regression relationships over time*, Math. Oper. Statist., Ser. Statist., 13, pp. 21–32.

—— (1982c), *Tests for change points based on recursive U-statistics*, Sequen. Anal., 1, pp. 263–284.

—— (1983a), *Some recursive residual rank tests for change points*, in Recent Advances in Statistics: Papers in Honor of Hermann Chernoff's Sixtieth Birthday, M. H. Rizvi, J. Rustagi and D. Siegmund eds, Academic Press, New York, pp. 371–391.

—— (1983b), *Recursive M-tests for the change points*, Internat. Statist. Inst. Proc. (Madrid Conf.), pp. 206–209.

—— (1983c), *Sequential R-estimation of location in the general Behrens–Fisher model*, Sequen. Anal., 2, pp. 311–335.

—— (1984a), *The Cox regression model, random censoring and locally optimal rank tests*, J. Statist. Plan. Inf., 9, pp. 355–366.

—— (1984b), *Invariance principles for U-statistics and von Mises' functionals in the non-i.d. case*. Sankhyā. Ser. A, 46, pp. 416–425.

—— (1984c), *Jackknifiing L-estimators: Affine structure and asymptotics*, Sankhyā, Ser. A, 46, pp. 207–29.

—— (1984d), *A James-Stein detour of U-statistics*, Comm. Statist., A13, pp. 2725–2747.

—— (1984e), *Recursive M-tests for the constancy of multivariate regression relationships over time*, Inst. Statist., Univ. North Carolina, Chapel Hill, Mimeo Rep. 1458; Sequen. Anal. 3, pp. 191–211.

—— (1984f), *On sequential nonparametric estimation of multivariate location*, Proc. Third Prague Conf. on Asymp. Math., pp. 119–130.

—— (1984g), *Subhypotheses testing against restricted alternatives for the Cox regression model*, J. Statist. Plan. Inf., 10, pp. 31–42.

—— (1984h), *Nonparametric testing against restricted alternatives under progressive censoring*, Inst. Statist., Univ. North Carolina, Mimeo Report. 1473.

P. K. SEN AND M. GHOSH (1971), *On bounded length sequential confidence intervals based on one-sample rank order statistics*, Ann. Math. Statist., 42, pp. 189–203.

—— (1974), *On sequential rank tests for location*, Ann. Statist., 2, pp. 540–552.

—— (1980), *On the Pitman-efficiency of sequential tests*, Calcutta Statist. Assoc. Bull., 29, pp. 65–72.

—— (1981), *Sequential point estimation of estimable parameters based on U-statistics*, Sankhyā. Ser. A, 43, pp. 331–344.

P. K. SEN AND A. K. M. E. SALEH (1984), *Nonparametric shrinkage estimators of location in a multivariate simple regression model*, Proc. 4th Pannonian Symp. Math. Statist., to appear.

—— (1985), *On some shrinkage estimators of multivariate location*, Ann. Statist., 13, pp. 272–281.

A. N. SHIRYAEV (1963), *On optimum methods in quickest detection problems*, Theor. Veroy. Prim.,
    8, pp. 26–51; Theory Prob. Appl., 8, pp. 22–46.
——— (1978), *Optimal Stopping Rules*, Springer-Verlag, New York.
A. N. SINHA AND P. K. SEN (1979a), *Progressively censored tests for clinical experiments and life
    testing problems based on weighted empirical distributions*, Comm. Statist. Theor. Meth., A,
    8, pp. 819–841.
——— (1979b), *Progressively censored tests for multiple regression based on weighted empirical
    distributions*, Calcutta Statist. Assoc. Bull., 28, pp. 57–82.
——— (1982), *Tests based on empirical processes for progressive censoring schemes with staggering
    entry and random withdrawal*. Sankhyā, Ser. B, 44, pp. 1–18.
——— (1983), *Staggering entry, random withdrawal and progressive censoring schemes: Some
    nonparametric procedures*, Proc. Golden Jubilee Conf. I SI, Calcutta, pp. 531–547.
E. V. SLUD (1982), *Consistency and efficiency of inferences with the partial likelihood*, Biometrika,
    69, pp. 547–552.
Y. C. SO AND P. K. SEN (1981), *Repeated significance tests based on likelihood ratio statistics*,
    Comm. Statist., Theor. Meth. A, 10, pp. 2149–2176.
——— (1982a), *M-estimators based repeated significance tests for one-way ANOVA with adapta-
    tions to multiple comparisons*, Sequen. Anal., 1, pp. 101–119.
——— (1982b), *Nonparametric repeated significance tests for one-way ANOVA with adaptations to
    multiple comparisons*, J. Statist. Plan. Inf., 7, pp. 83–96.
R. N. SPROULE (1974), *Asymptotic properties of U-statistics*, Trans. Amer. Math. Soc., 199, pp.
    55–64.
C. STEIN (1945), *A two-sample test for a linear hypothesis whose power is independent of σ*, Ann.
    Math. Statist., 16, pp. 243–258.
A. A. TSIATIS (1981a), *A large sample study of Cox's regression model*, Ann. Statist., 9, pp. 98–108.
——— (1981b), *The asymptotic distribution of the efficient scores test for the proportional hazard
    model calculated over time*, Biometrika, 68, pp. 311–315.
A. WALD (1947), *Sequential Analysis*, John Wiley, New York.
A. WALD AND J. WOLFOWITZ (1948), *Optimal character of the sequential probability ratio test*, Ann.
    Math. Statist., 19, pp. 326–339.
S. WATANABE (1964), *On discontinuous additive functionals and Lévy measures of a Markov
    process*, Japanese J. Math., 34, pp. 53–70.
M. WOODROOFE (1982), *Nonlinear Renewal Theory in Sequential Analysis*, CBMS Regional
    Conference Series in Applied Mathematics 39, Society for Industrial and Applied
    Mathematics, Philadelphia,

Index

99

2)